19TH CENTURY
FEMALE EXPLORERS

Some historical sources used in this book include outdated language and negative attitudes towards people and/or cultures. They do not reflect the opinions of Pen & Sword or the author.

19TH CENTURY FEMALE EXPLORERS

CAROLINE ROOPE

PEN & SWORD HISTORY

AN IMPRINT OF PEN & SWORD BOOKS LTD.
YORKSHIRE – PHILADELPHIA

First published in Great Britain in 2023 by
PEN AND SWORD HISTORY
An imprint of
Pen & Sword Books Ltd
Yorkshire – Philadelphia

Copyright © Caroline Roope, 2023

ISBN 978 1 39900 686 6

The right of Caroline Roope to be identified as Author of this work has been asserted by her in accordance with the Copyright, Designs and Patents Act 1988.

A CIP catalogue record for this book is available from the British Library.

All rights reserved. No part of this book may be reproduced or transmitted in any form or by any means, electronic or mechanical including photocopying, recording or by any information storage and retrieval system, without permission from the Publisher in writing.

Typeset in Times New Roman 12/16 by
SJmagic DESIGN SERVICES, India.
Printed and bound in the UK by CPI Group (UK) Ltd.

Pen & Sword Books Limited incorporates the imprints of Atlas, Archaeology, Aviation, Discovery, Family History, Fiction, History, Maritime, Military, Military Classics, Politics, Select, Transport, True Crime, Air World, Frontline Publishing, Leo Cooper, Remember When, Seaforth Publishing, The Praetorian Press, Wharncliffe Local History, Wharncliffe Transport, Wharncliffe True Crime and White Owl.

For a complete list of Pen & Sword titles please contact
PEN & SWORD BOOKS LIMITED
George House, Units 12 & 13, Beevor Street, Off Pontefract Road,
Barnsley, South Yorkshire, S71 1HN, England
E-mail: enquiries@pen-and-sword.co.uk
Website: www.pen-and-sword.co.uk

or

PEN AND SWORD BOOKS
1950 Lawrence Rd, Havertown, PA 19083, USA
E-mail: uspen-and-sword@casematepublishers.com
Website: www.penandswordbooks.com

Contents

Acknowledgements	vii
Preface	viii
Introduction: Beyond the garden gate	x
Prologue: The voyage out	xviii

Part 1	For the love of God Susie Rijnhart Annie Boyle Hore Annie Royle Taylor	1
Part 2	For the love of a good (or bad) man Margaret Fountaine Lady Jane Digby Lady Hester Stanhope Lady Jane Franklin Isabella Bird	24
Part 3	For science and antiquities Mary Brodrick Marianne North	68
Part 4	For fame, fortune and newspapers Annie Londonderry Elizabeth 'Nellie' Bly Martha Black	80

Part 5	For Duty	111
	Susanna Moodie	
	Catharine Parr Traill	
	Eliza Bradley	
	Cara Edgeworth David	
	Isobel Gill	
	Florence Baker	
	Mary Livingstone	
Part 6	For the sheer fun of it	146
	Ida Pfeiffer	
	Aimée Crocker	

Epilogue: The journey home	177
Notes	181
Bibliography	202
Index	211

Acknowledgements

A heartfelt thank you must go to our explorer ladies themselves. If they hadn't left behind such a vast catalogue of narratives detailing most, if not all, of their adventures, this book would have been impossible to write. If they trod where others feared to tread, then we are all the richer in our knowledge and understanding because of it. Their stories are fascinating, and this book is only a snapshot. I urge everyone to read their original narratives – not only to help resuscitate these (in some cases) long forgotten travelogues from gathering dust in the corner of a library, but for the sheer delight of listening to their voices resonate from the page.

To the authors who have studied this subject previously, and whose work I have drawn upon and, in some cases, quoted, thank you – not only for happily granting permission for me to use your work, but for recognising the achievements of these remarkable ladies and helping to share their stories with a wider audience – something I hope I have achieved myself in this book.

Thank you also to my long-suffering husband for his enduring support – if I were shipwrecked, you'd make a very handy flotation device.

And finally, I would like to dedicate this work to my sister, Katie, the bravest and strongest woman I know. You may not have travelled far in comparison to some of these ladies, but you have navigated your way through the last ten years like a pro. If I were faced with a tribe of angry locals, I'd want you there to talk us out of trouble.

To all the strong women out there – this book is for you.

Preface

They climbed pyramids in their long frocks, paddled canoes through the rapids of West Africa, and rode across the Rocky Mountains on horseback. Whilst nineteenth-century Britain was establishing itself as a powerhouse of industry and innovation, a handful of plucky women were forging their own paths abroad – often armed with machetes, pistols and more than a little derring-do.

Travel and exploration in the nineteenth-century was not for the faint-hearted. There was the constant threat of disease and illness, as well as the danger of coping with extreme climates. Journeys were perilous; and transport, if there was any, was basic. Despite this, the century saw women donning their thickest skirts and toughest boots and taking a surprisingly big step out of their domestic sphere and into the unknown. Their devil-may-care attitude and sheer feats of courage should have assured their place in the annals of history - yet these women remain remarkably elusive. With the exception of Gertrude Bell, whose role in Middle Eastern politics seems to have cemented her place in collective memory, the likes of Margaret Fountaine, the lepidopterist extraordinaire who was hoping to catch more than just butterflies in her net; Isabella Bird, the first female fellow of the Royal Geographical Society; and Lady Hester Stanhope, trouser and sabre-wearing socialite *par excellence*; have all but disappeared from view.

Their narratives and travelogues, of which there are many, have regularly been dismissed as eccentric ramblings – or worse, romanticised accounts of their travels around the Empire. As women, they were expected to stay at home and attend to their domestic responsibilities. Marriage and motherhood were the paradigm by which women lived their lives, but somehow these pioneering women found a way to burst out of their stuffy drawing rooms, transcend the confines of their sex and reach out to the rest of the world.

These adventure-hungry women didn't set out to consciously break the mould. Like their modern-day counterparts, they just wanted to do something exciting and experience something different. They weren't actively trying to score a point for feminism, and they certainly didn't consider themselves trailblazers. The authenticity of their voices lies in the fact that they chose to travel and did so of their own volition - satisfying only themselves. Their accounts are free of the rhetoric that would appease a wealthy patron or uphold a professional reputation. In the words of Mary Kingsley just before she embarked on her first trip to West Africa in 1893; 'My mind was set on going and I had to go.'

And go they did. To the furthest reaches of the earth, finding freedom and self-expression along the way. In a century that saw the rise of the suffrage movement and a growing awareness of women's rights, this book uncovers their journeys – both personal and physical; and in some cases, political too – and the challenges they faced. It also celebrates the achievements of these extraordinary women – women who chose to turn their back on convention and risk society's disapproving looks for the sake of living life by their own rules. Women whose sense of determination and adventurous spirit expanded not only their own horizons but ours too. Women whose voices still resonate centuries later.

Introduction

Beyond the garden gate

> I craved to go beyond the garden gate, to follow the road that passed it by, and to set out for the unknown.[1]
>
> Alexandra David-Neel, 1927

As a child runaway, it would take more than a garden gate to stop Alexandra David-Neel. After absconding from a family holiday in the Netherlands, the fifteen-year-old David-Neel made it as far as England, having previously cut her explorers teeth at the tender age of five when she attempted to escape the noise of her new baby brother by walking out through the garden gate that was installed to stop her doing just that.

Doors, gates, fences, hedges - all barriers that provide a tantalising challenge to the would-be explorer. If one could conquer a lowly gate, surely the vast and shifting sands of the Arabian desert, or the fathomless depths of the Atlantic Ocean could be overcome with similar ease. Such was, and still is, the mentality of some of the world's greatest explorers. A desire to see what is over the horizon or 'just around the corner' is familiar to all of us, yet in the explorer it is often akin to an irrepressible urge; a preoccupation bordering on the obsessive. The desire to go beyond – not just geographically but in many cases physically, mentally and spiritually – to the very limits of civilisation and human endurance are embedded in the genetic makeup of even the earliest globetrotters.

We take for granted our twenty-first century ability to just 'go' when and wherever we please – unless war or global politics dictate otherwise. But for the nineteenth century female explorer 'going' anywhere meant negotiating the choppy seas of society's expectations before a gangplank could be walked up or a sail hoisted. As empires began to expand in the early nineteenth century, a woman's horizons shrunk. Her place was at home, unless her husband had the fortune - or misfortune, in some cases - to be posted to a colonial outpost or filled with the sort of missionary zeal that meant a pilgrimage was on the cards. Even these women were seen as an appendage to the journey - tagging along out of necessity rather than by choice. Denied the freedom to roam at will, women invariably turned inwards – into their own emotions and the physical and mental well-being of their husbands and children.

According to Mrs Ellis – who took it upon herself to provide moral guidance to English ladies through several 'handy' guides - women were expected to show the 'strength and dignity of character, a power of usefulness, and a capability of doing good,'[2] that their eighteenth-century predecessors apparently had in abundance. These women seldom travelled abroad, she writes, because 'Their sphere of action was at their own firesides, and the world in which they moved was one where pleasure of the highest, purest order, naturally and necessarily arises out of acts of duty faithfully performed.'[3] Woe betide those who had ambitions beyond the front door or wanted to travel for their own reasons. Such women were an oddity. At best they were cast as eccentric spinsters with too much money and not enough sense, at worst they were unladylike, over-ambitious and clearly had delusions beyond the rest of their sex. For the moralistic Mrs Ellis, the problem was their inability to accept their destiny – as a dutiful daughter, wife and mother. These agitators were 'distinguished by a morbid listlessness of mind and body, except when under the influence of stimulus, a constant pining for excitement, and an eagerness to escape from everything like practical and individual duty.'[4] Shame on them.

'A Lady an explorer? A traveller in skirts? The notion's just a trifle too seraphic: Let them stay and mind the babies, or hem our ragged shirts; But they mustn't, can't and shan't be geographic,'[5] postulated a journalist in *Punch* magazine in 1893. His poetic outburst was prompted by the shocking news that fifteen ladies had finally, after a mere forty-five years of discussion, been elected as Fellows of the Royal Geographical Society. Lagging behind the Royal Scottish Geographical Society, who had been admitting women since its inception in 1884, the RGS declared that the women had to be 'well-qualified,' to which one Fellow enquired whether that meant in beauty or in science? Another decided that it would spell the beginning of the end for the Learned Society, which would undoubtedly now become a Garden Party Society – which was ironic given that the original Fellows had previously shared rooms with the Horticultural Society in Regent Street. With not a hint of sour grapes, George Nathaniel Curzon, recently returned from his own adventures in Asia, voiced his opinion on the matter most vociferously: 'Their sex and training render them equally unfitted for exploration, and the genus of professional female globe-trotters with which America has lately familiarized us is one of the horrors of the latter end of the nineteenth century.' Not to be put off, one of the recently elected Fellows, Kate Marsden, applied for a ticket to the RGS's Annual Dinner in 1894. She was refused on the grounds that she would be 'the only lady among 200, nearly all of them smoking.'[6] Given Miss Marsden's recent mission to nurse the leper colonies hidden in the forests of deepest, darkest Siberia, from which she returned on horseback with an escort of Cossack soldiers, one would assume this wouldn't have unduly troubled her.

With such hostility it is remarkable that women persisted in their endeavours to travel outside the drawing room, let alone their home country. But they did – and we, as armchair explorers, are all the better off for it, particularly as they left behind such remarkable accounts of their travels. Often scoffed at by their male contemporaries - after all women didn't write books - their letters, diaries, journals and travelogues provide us, the reader, with the means to voyage along with them. They can be funny, as in the case of Mary Kingsley, who on finding herself falling to the bottom of an animal trap stated without a hint of irony that, 'It is at these moments you realise the blessings of a good thick skirt.'[7] And they often hint at their innermost thoughts; 'Palled for the moment with civilisation and its surroundings, I wanted to escape somewhere, where I might be as far removed from them as possible,'[8] wrote Lady Florence Dixie in 1880 about her decision to explore Patagonia. They can be brutal in their honesty, as Emily Lowe demonstrates so succinctly in *Unprotected Females in Norway* (1857): 'The only use of a gentleman in travelling is to look after the luggage, and we take care to have no luggage.'[9] And perhaps most importantly they give us their impressions – not just of the places they went, but the people they met: 'Women and children – who appear to us wonderfully fair after the natives of the plains – are employed in plucking the leaf, which they throw into long upright baskets, the men being reserved for the more laborious operations of hoeing, planting, etc. We pass groups of patient women thus busily occupied, whilst wee babies, from ten days old and upwards, in shallow baskets made to fit them, lie speckled about the ground; placed by maternal solicitude beneath the scanty shadow of the tea bushes, each looking like a little Moses, minus the bulrushes,'[10] wrote Nina Elizabeth Mazuchelli in *The Indian Alps and How We Crossed Them* (1876). It is this cultural commentary that sets the female explorer's account apart from her male counterparts. As author Mary Morris explains, 'women move through the world differently than men. The constraints and perils, the perceptions and complex emotions women journey with are different from those of men.'[11] While men travelled with the gathering of data in mind – often to please a wealthy patron or to forge a reputation - nineteenth-century female explorers, for the most part, had no such obligations. In this respect they were free of at least one set of shackles and could instead focus on recording the richness of life. From a description of what they were offered to eat at a roadside inn (in Mazuchelli's case, a plate of grilled fowl and chupattee known locally as *sudden death*) to the people they met on their travels (for May French Sheldon, a village 'in an absolute state of nudity'[12]) - their reflections on their adventures are as deep and varied as the reasons they chose to go in the first place.

So, what were those reasons? And at what point did women stop being a mere 'accompanist' on a journey and start to strike out on their own? Although I have attempted to categorise these extraordinary women based on why they chose to go exploring, the task is a thankless one since the women in question do their absolute best to avoid being pigeon-holed. Thus, my categories must be taken in the broadest possible sense – inevitably there are crossovers and as each explorer meanders along on her journey, her path will often cross with that of another. So, whilst Margaret Fountaine's extensive butterfly hunting expeditions are discussed in the section covering those who travelled for love, she could just have easily been included in the section detailing those who explored for the sheer fun of it. Likewise, Canadian explorer Susie Rijnhart's decision to travel was forged out of love and devotion to God - but also with a sense of duty towards her husband that led her to continue the missionary work that they had begun together after his tragic death. Her journey was one of grief as well as love; but also, hope and renewal in the face of unimaginable hardship.

Their external journeys often coincided with an internal one – as they explored their surroundings, they were, consciously or not, exploring their inner landscapes, their beliefs, their values, and the limits of their own endurance. They crossed not just physical borders, but the borders of their own psyche. American writer Alice Walker once reflected, 'The most foreign country is within,' and that is certainly true of many nineteenth-century female explorers. Their writings are laced with a dialogue of what is happening both without and within – reflections that can, at times, make for painful reading. Travelling through North Africa at the turn of the twentieth-century, Isabelle Eberhardt's diary is an uncomfortable monologue of a life in tatters. Born into Genevan aristocracy, Eberhardt embraced a nomadic existence – revelling in the freedom she found from the obligations imposed by her class and sex in Europe. As a Muslim convert, she dressed as a man to explore the Arab world, but her conflicting personas were her downfall. Her conversion to Islam ostracised her from the expatriate European society in the Arab states, and Arab custom dictated that as a 'man' she was to have no formal contact with women. Her diary charts her spiral into drugs and homelessness as she became increasingly isolated from the world around her. 'I was right to account for the wretched way I live down here as mere eccentricity: that way, it is not too obvious that I am in fact destitute. I have begun to make a point of going to people's houses to *eat*, for the sole purpose of keeping fit, something that would have been *anathema* in the old days, like the other thing I have been doing lately, namely going to see marabouts [Muslim religious leaders], just to beg them for money,'[13] she wrote. Eberhardt died at the age of twenty-eight in a flash flood, with the profound belief that 'The human body is nothing, the human soul is all.'[14]

All the women discussed here voyaged not just to unfamiliar lands, but across the invisible boundaries of a patriarchal society and the customary conventions of their sex. But freedom could also be a dangerous thing. Fear of attack was an ever-present danger to the female explorer, but other perils were, and still are, more subtle, such as harassment and unwanted attention. British travel author, Annette Meakin, was quickly advised to procure herself a handgun when exploring Siberia at the turn of the nineteenth century. Having convinced her mother to join her on her trip on the Trans-Siberian railway, and after reducing their luggage to a valise, holdall, and the all-important tea basket, a fellow lady-traveller was horrified to discover the pair had no gun. 'This lady thought us bold, if not rash, to travel without a revolver. 'I always carry two,' she said.'[15] Attitudes towards the 'weaker' sex and their ability to show bravery in the face of danger were slow to change and deferring to the nearest man in such situations was still the recommended course of action. In *Hints to Lady Travellers: At Home and Abroad,* the Victorian version of a *Rough Guide to...,* writer and cyclist Lillias Campbell Davidson suggests, in rather blasé fashion, that, 'It is so much an instinct with the stronger sex to protect and look after the weaker, that in all cases of the sort, if there is a man at the head of affairs, he had better be left to manage matters without the hampering interference of feminine physical weakness.'[16] This, of course, was only helpful if the said man was not the aggressor. Her fellow lady-cyclist, twenty-four-year-old Annie Londonderry, knew a thing or two about tackling bandits. On her around the world cycling trip, which she started in 1894, her pearl-handled pistol was put to effective use when, according to Londonderry – who was prone to flights of fancy – she was accosted at gunpoint by bandits. That Londonderry would later stage manage a photo of the alleged holdup to use as a lecture slide did nothing to diminish the veracity of her claim – such was the power of her own hype, and her hunger for fame and fortune. Annie Londonderry sits in a class of her own among these women, being the only one whose motive for exploring allies so clearly with a desire to profit from her adventures. Ever the glory hunter, Londonderry would no doubt have revelled in her distinction from her less sensational travelling sisters.

As we read, we are invited to travel with them – across foreign lands aplenty, but also into their inner world, and experience the globe through a nineteenth-century lens. We are also invited to admire their achievements, of which there are many - but this is not a book of 'firsts,' or records broken. Perish the thought since they themselves saw no glory in their accomplishments. 'At the time it seemed throughout an easy, commonplace affair which anybody could have accomplished,'[17] wrote Jane Ellen Duncan, of crossing the Chang La mountain pass, despite the threat of severe altitude sickness. 'I have no gift of fine writing to cast a glamour over it and make it seem the tremendous achievement it was not.'[18] Despite their modesty, their journeys speak for themselves, whether they were the third or fourth woman to climb Mont Blanc, or they were bringing up a gaggle of children whilst trekking through the African bush.

What unites them all is a sense of wonderment. For some, their only regret is that words cannot do it justice;

> Silently we glided away from the torchlight into the apparently impenetrable darkness, but the heavens, of

which we saw a patch now and then, were ablaze with stars, and ere long the forms of trees above and around us became tolerably distinct…Sheet lightning, very blue, revealed at intervals the strong stream swirling past under a canopy of trees falling and erect, with straight stems one hundred and fifty feet high probably, surmounted by crowns of drooping branches; palms with their graceful plumage; lianas hanging, looping, twisting – their orange fruitage hanging over our heads…Oh that you could see it all! It is wonderful; no words could describe it, far less mine. Mr Darwin says so truly that a visit to the tropics (and such tropics) is like a visit to a new planet. This new wonder-world, so enchanting, tantalising, intoxicating, makes me despair, for I cannot make you see what I am seeing![19]

Prologue

The voyage out

'Of the voyage to Hong Kong little need be said. The 'Volga' is a miserable steamer, with no place to sit in, and nothing to sit on but the benches by the dinner-table in the dismal saloon… The ship was damp, dark, dirty, old and cold. She was not warmed by steam, and the fire could not be lighted because of a smoky chimney. There were no lamps, and the sparse candles were obviously grudged. The stewards were dirty and desponding, the serving inhospitable, the cooking dirty and greasy, the food scanty, the table-linen frowsy.'

Isabella L. Bird, 1883

To explore somewhere new, one needs to get there – hopefully in one piece. Whether by train, horse, bicycle, sledge, ship, camel, donkey or on foot, plans must be drawn up, maps purchased, and luggage packed. Loved ones must be informed and goodbyes said. Woe betide those who travel unprepared – leaving one's destiny to chance could quite literally be the difference between life and death.

Fortunately for thirty-one-year-old Ella Constance Sykes, she had the luxury of ten days to prepare for her new life abroad. In 1894, the Foreign Office commissioned her brother Percy to establish consulates in the Persian districts of Kerman and Baluchistan. Spotting an opportunity to have, in her own words, a 'quantum of adventure,'[1] and bored of being 'civilised all my days', Sykes was ready to throw off her late-Victorian shackles in favour of 'a sense of freedom and expansion which quickened the blood and made the pulse beat high.' Exotic ecstasies were clearly high on her agenda, and by tagging along with her little brother, she hoped to find them in Persia. But first she had to tackle the onerous task of packing a substantial quantity of luggage. 'It was necessary to take sufficient clothing to last for over a year, to buy a complete camp equipment, and to lay in furniture, linen, glass and crockery for the new establishment at Kerman, and, what was perhaps more important than anything else, to engage a maid.'[2] Sykes's shopping list was no mean feat in a pre-Amazon world, but of all the baggage a globe trotter could expect to take with her, the correct clothing was one of the most imperative – particularly if they had Mrs Ellis's warning that 'A careless or slatternly woman...is one of the most repulsive objects in creation,'[3] ringing in their ears. But lady explorers had larger, more practical issues to contend with than whether they had a tatty hem.

Get an outfit wrong and not only could it hinder a woman's ability to actually *get* anywhere, but it could also cost her her life. Mary Kingsley's 'good thick skirt,' came into its own in the most dramatic way, but other clothing related perils were harder to manage – particularly if they involved the weather.

When English nurse and friend-to-the-Siberian-lepers, Kate Marsden, tried to board a sledge to Yakutsk in 1890 it took the efforts of three Russian policemen to help hoist her onto it – such was the stiffness and bulk of her numerous layers. In her own words it was 'a decided burden in more ways than one.'[4] Marsden's outfit included 'a whole outfit of Jaeger garments'[5] - a flannel lined 'body,' a thickly-wadded eiderdown ulster with a fur collar to shield the neck and face, a floor-length sheepskin, and over that, a fur coat of reindeer skin. Underneath she wore thick Jaeger stockings, covered with a further pair of gentleman's hunting stockings. For her feet she wore a pair of knee-high Russian boots made of felt and covered those with a pair of brown felt *valenki* (traditional Russian winter footwear). 'It was not surprising that, when thus accoutred, broadened, and lengthened by a great many inches, I failed to recognise K. M. in the looking glass,'[6] she remarked. A black and white photograph of Miss Marsden in her travelling attire accompanies the description she gives in the account of her travels, *On Sledge and Horseback to Outcast Siberian Lepers*. It is easy to see why bending was an issue.

Clothing, however, was the least of Marsden's worries. Born into a comfortable London household in 1859, by eighteen she would find herself having to make her own living. Her calling was nursing but although her first foray into medicine led her to her local hospital, Marsden was destined for nursing on an international stage, and she promptly found herself in a more challenging environment – tending casualties of the Russo-Turkish war in Bulgaria. The experience changed her life, for it was there that she first encountered the sufferings of lepers, which led her to plan her 'Mission' – to alleviate and find a cure for leprosy.

Taken at face value, the title of Marsden's account, *On Sledge and Horseback to Outcast Siberian Lepers* sounds far-fetched. Firstly, she was – *gasp* - a thirty-one-year-old spinster, which was akin to being a non-entity in the eyes of the Victorian world. Secondly, she was in poor health herself, suffering from the oft-quoted nineteenth-century twin maladies of consumption and nervous anxiety. Her 'Mission' involved riding two-thousand miles across one of the most inhospitable regions on earth to find a magic herb that grew deep in the Siberian Forest – a herb that she believed would cure leprosy. That she would later encounter disbelief at her adventures is understandable. But riding on a sledge and horseback to outcast Siberian lepers was exactly what she did. It was a grim and thankless task, particularly as the Russian government took a dim view of her meddling with their lepers. Fortunately, she kept her humour throughout, and her writing is peppered with amusing anecdotes. On her very nearly abortive attempt to get in the sledge, she wrote drily that 'Three muscular policemen attempted to lift me gently into the sledge; but their combined strength was futile under the load. So, they had me set on the ground again.' Her trial did not end there, however. 'Then I attempted in a kind of majestic, contemptuous way to mount without assistance; but, alas! My knees would *not* bend.' She was eventually hauled in, her sense of pride now forgotten, and the 'undignified operation,' thankfully over. Her only regret was that once on the sledge she couldn't bend herself in half to perform a bow of thanks to those who had assisted her.

Both Isabella Bird and Annie Londonderry found their initial clothing choices a hindrance rather than a help, and both would resort to commissioning entirely new and - for the time – sensational outfits. Isabella Bird, one of Victorian Britain's most well-known and well-travelled explorers, was extremely keen to dispense of her standard cumbersome skirts and in the spirit of her travelling mantra that 'Travellers are privileged to do the most improper things with perfect propriety,'[7] she decided that since she was thousands of miles away from the disapproving stares of Victorian society, that her expedition through the Rocky Mountains in 1873 would be a much more enjoyable experience if she could sit astride her horse, rather than ride side-saddle. To this end, she took to the mountains in a special 'riding dress' which comprised a 'half-fitting jacket, a skirt reaching to the ankles, and full Turkish trousers gathered into frills which fall over the boots.'[8] Such a departure from the normal mode of feminine attire meant that not only could Bird truly indulge in the 'up-to-anything free-legged air'[9] that she advocated (quite literally 'free-legged,' in her case) but also guaranteed that she secured the condemnation of the press. *The Times* reported her as having worn 'masculine habiliments', much to her chagrin, and she retorted by insisting that her outfit was 'thoroughly feminine.' As the grand dame of the travelling world, and the first woman to address a meeting of the Royal Geographical Society, Bird was not to be crossed.

Annie Londonderry's challenges were a little more ergonomic. Not only did she need to sit astride something – in her case, a bicycle – but she also needed to be able to move her legs up and down to power it. Born Annie Cohen in 1870, Londonderry's notoriety as the first female to bicycle around the world coincided with the cycling craze of the late nineteenth-century. As a Jewish Latvian immigrant to the United States, Annie Cohen (who, confusingly, got married and became Annie Kopchovsky before adopting the name 'Londonderry') was an unlikely candidate for such a feat. In her version of events, two wealthy Bostonians wagered $20,000 against $10,000 that no woman could travel around the world by bicycle in fifteen months and earn $5000. Hearing of the wager, Annie volunteered to, quite literally, 'spin' her way around the world. The wager marked the beginning not only of Annie's adventures, but also some very tall tales. Hunting tigers with royalty, being sent to a Japanese prison with a bullet wound, exaggerated accidents and injuries - all of this and more made it into Annie's narrative. Fortunately for Annie, her 'exploits' brought her more column-inches, which made her more popular with a gullible public and, more importantly, her numerous sponsors.

Notwithstanding the fact that Annie had never ridden a bike before, she was also married with three small children. What would she do with them? Her husband Max couldn't be expected to look after them. Or could he? Annie herself was emphatic about her life choices, stating quite unreservedly that 'I didn't want to spend my life at home with a baby under my apron every year.'[10] When *The World* suggested to Annie 'that she might carry her children along with her on a bicycle built for four,'[11] she answered that 'she had enough troubles of her own' and the children were not mentioned again.

Annie's 'troubles' began with her skirts. The first leg of her journey from Boston to New York in the summer of 1894 was hard going in her usual attire, which consisted of a long blue serge skirt, corset and high-collared blouse. In Annie's words, it was 'getting in the way whenever I tried to make any speed.'[12] Her new suit, which was whipped up for her before she left New York, was made of 'dark blue henrietta cloth' with a 'comfortable blouse waist.' She accessorized the look with 'rubber sole shoes, no corsets and a jaunty cap to match her suit…When the riding is hard and the skirt seems in the way it is easily pinned up.' Her unadorned but practical attire would have pleased Lillias Campbell Davidson who was insistent that for cycling 'All flowers, bright ribbons, feathers etc. are in the worst possible taste, and should be entirely avoided.'[13] Very sensible.

Londonderry became part of a wider movement concerned with dress reform and gender equality. The bicycle was both a literal and symbolic vehicle for social change at the end of the nineteenth-century, and Annie's globetrotting enterprise provided an opportunity to demonstrate women's new-found freedom. There was one slight problem, however. The ultimate instrument of feminism had no brakes, which was ironic given they were also known in the nineteenth century as 'safety machines.' They weren't just representative of emancipation they were also a death-trap. Annie would need those rubber soled shoes for more than just practicality.

Unlike Annie Londonderry, some ladies had their choice of clothing thrust on them by accident rather than by design. When Lady Hester Stanhope was shipwrecked near the island of Rhodes in 1811, she lost all her luggage and was forced to adopt the clothing sported by the locals. As a free-spirited and open-minded lady of the world, Lady Hester embraced her change in circumstances with pragmatism: 'To dress as a Turkish woman would not do, because I must not be seen to speak to a man; therefore I have nothing left for it but to dress as a Turk.'[14] For Hester, this meant donning several layers of silk and cotton shirts, a waistcoat, long breeches and a sash around her waist. She also tucked a pistol and knife into her sash for good measure and put a turban on her head. She was evidently more pleased with her look than Kate Marsden was with hers for she boasted in a letter that 'I can assure you that if I ever looked well in anything it is in the Asiatic dress.'[15] For Stanhope, it marked a life-long passion for dressing in Asian clothing, which was entirely in keeping with her splendidly eccentric approach to life.

Annie Londonderry and Lady Hester Stanhope weren't the only female explorers dicing with death when they chose their mode of transport. Nervous traveller, Janet Robertson was less than thrilled by her passage to Italy along 'the wide waste of waters,' in which she experienced 'rain, wind, and raging storms, which momentarily threatened to send the vessel to the bottom.'[16]

For some, luxurious habits die hard – even in the African wilderness. American explorer and ethnographer, May French Sheldon liked to travel in style. Having survived the initial journey to Africa in 1891, the bombastic Sheldon – whose entourage included 153 porters, a personal masseuse, a portable bath, Parisian silk dresses and a blonde wig – journeyed into the African Interior in an enormous wicker palanquin strewn with luxury furnishings. She nicknamed the palanquin her 'white elephant.'[17] Sheldon may have reeked of the ultimate white traveller abroad – her native male porters called her 'Bebe Bwana' (woman master) – but behind the imperialist façade lurked an extremely talented and courageous scientist with a genuine interest in observing and documenting other cultures. Even getting hit in the face by a branch and having a thorn impaled in her eyeball didn't stop her: 'After the shock had abated, my eye was bandaged, and on we marched. One does not stop for an eye or a limb or a life in Africa; one is ever impelled to proceed...'[18] One night, she almost lost her beloved palanquin to a giant python, awakening to find the terrifying reptile attempting to crush the wicker chair with her inside. 'I became aware, through the atmospheric conditions, that a cold, clammy, moving object was above me, in truth almost touching me, on the top of my Palanquin, the rattans of which were cracking as if under the pressure of a mangle,'[19] she later recounted. 'My blood fairly seemed to congeal in my veins at the spectacle: it was an enormous python, about fifteen feet long, which had coiled around the top of the Palanquin, and at that moment was ramping and thrusting its head out, searching for some attainable projection around which to coil it's great, shiny, loathsome length of a body...I am ashamed to confess it was the supreme fear of my life, and almost paralyzed me.'[20] Sheldon would also survive an episode of near-drowning - her palanquin very nearly provided her with a watery coffin when she was accidentally deposited into a 'swollen torrent,' along with several of her porters.

Sheldon found herself 'head down and completely submerged in thick yellow water, in jeopardy of drowning.'[21] She was hauled out but to make matters worse, she was dropped in again during the rescue attempt and suffered a severe injury to her spine. She was eventually carried 'limp and helpless, as I thought permanently disabled, up the bank.'[22] Fortunately, Sheldon's injuries were not serious and despite her harrowing experience, she happily returned to Africa in 1894 to see the Belgian Congo.

Ella Sykes managed to narrowly avoid death-by-horse-and-carriage on her way to the Persian town of Resht in 1894. Six miles along a primitive 'road' in a set of 'small and rickety' horse drawn carriages, was enough to test anyone's nerve, let alone the Western sister of a diplomat, whose travelling experience didn't extend beyond continental Europe. She wrote: 'My experiences in Constantinople had only prepared me in a very small degree for the inexpressible badness of this road, streaming with mud and water as it was. We bumped in and out of holes, were nearly overturned at exceptionally bad bits of the road, and my companion and I had to cling tightly to one another to save ourselves from being thrown out. Every moment I felt that our dilapidated vehicle must perforce come to pieces, or that the insecurely fastened wheels must succumb to the repeated shocks as we jolted painfully along.'[23] She was eventually rewarded for her perseverance with 'the pretty town of Resht embowered in autumnal-tinted trees, the red tiled roofs of its houses giving it at a distance much the look of an English country village.'[24]

And what of the England – or for that matter, America, Germany, France, Scotland and various other Western countries – they left behind? Did they chance a sentimental look back as they departed, or did they keep their eyes resolutely on the path ahead?

'It was at Liverpool,' wrote Rebecca Burlend of her imminent emigration to North America in 1831, 'when we had got our luggage to a boarding house and were waiting the departure of the vessel, that the throes of leaving England and all its endearments put our courage to a test the most severe.'[25] True, Burlend had never really wanted to go to America anyway; but with limited financial prospects in her native Yorkshire, she had little choice. 'As the wind was favourable we soon lost sight of the shore…and when it was finally announced that England was no longer visible, there was not a person in the ship who would not have heartily responded amen to the prayer 'God bless it'.'[26]

Her sentiments were shared by Mrs T. F. Hughes, the wife of a diplomat, who documented her wretched experiences in China and Formosa in the 1870s. 'Leaving home for the first time,' she wrote, 'with the prospect of a lengthened absence, is a sad and terrible undertaking; and when your destination lies at the other side of the globe…parting seems to be a sorrow almost too bitter to be borne.'[27] Rebecca Burlend and Mrs Hughes were in the minority, however; their experience of leaving home no doubt tinged with the regret of an unwilling traveller, and a grim acceptance of their fate - to do their duty by their husbands. They were reluctant accompanists; and miserable ones at that.

Well-meaning friends and family could also be relied upon to complicate a departure, either by looking sorrowful and weeping or by attempting to impart 'helpful' advice. Vienna-born Ida Pfeiffer had something of the sort when she was planning her pilgrimage to the Holy Land in 1842, writing: 'My friends and relations attempted in vain to turn me from my purpose by painting, in the most glowing colours, all the dangers and difficulties which await the traveller in those regions…to think of a woman's venturing alone, without protection of any kind, into the wide world, across sea and mountain and plain – it was quite preposterous. This was the opinion of my friends.'[28] Luckily, Pfeiffer didn't much care for the opinion of her friends, and neither did she fear death. Fully expecting to die on her 'preposterous' journey she drew up her will and organized all her worldly affairs so that her family would 'find every thing perfectly arranged,' should she expire during her travels.

Following a stultifying upbringing in which she had nursed an invalid mother and sickly brother, Mary Kingsley also had no qualms about the dangers she was likely to encounter. Being recently orphaned, and unmarried seemed to give her a slightly gung-ho attitude, and she appeared not in the least bit concerned for her own safety. She was finally in charge of her own destiny and intended to make the most of it, whether it put her in danger or not. Having cast off the shackles of duty she declared that she was 'Just off to skylark and enjoy myself in Africa.'

It was this irrepressible level of pragmatism that carried the female explorer through many a trial – the complexities and challenges of their domestic, pre-exploring lives providing an unpleasant but ideal environment to foster a hardening of the spirit. Scottish missionary and contemporary of Kingsley, Mary Slessor, spent her childhood avoiding the wrath of her alcoholic father. After a hard day at Baxter's textile works, during which she surreptitiously read the bible whilst operating a loom, she saw out her evenings dutifully sewing with her mother and 'waiting with sickening apprehension for the sound of uncertain footsteps on the stairs.'[29] When her father finally appeared, Mary often had little choice but to go out 'into the streets where she wandered in the dark, alone, sobbing out her misery.'[30] Whether preaching the word of God to the warring tribes of the Calabar provided a safer environment than the cold streets of Aberdeen in the mid-nineteenth century is debatable, but such was the life that Slessor was destined for.

Often their plans had been laid long before they voyaged out, as their difficult home lives forced them to retreat into the fertile land of their imagination – where they could ramble wherever they liked. Mary Kingsley's enquiring mind frequently led her to her father's library, where through his extensive book collection she lived 'in a great amusing world of my own other people did not know or care about.'[31] Mary Slessor's biographer, W. P. Livingstone believed her missionary calling to Africa was evident long before she set foot in a mangrove forest. He wrote: 'Often Mary played at keeping school, and it is interesting to note that the imaginary scholars she taught and admonished were always black,'[32] and that: 'The story of Calabar had impressed her imagination when a child, and all through the years her eyes had been fixed on the great struggle going on between the forces of light and darkness in the sphere of heathenism.'[33] Such sentiments jar when subjected to our twenty-first century scrutiny, yet it is impossible to disentangle the journeys and experiences of many of these remarkable women from the white imperialist legacy that informed their decisions. Inspired by tales of 'The Empire,' the crossing of frontiers and the formation of colonies of which their ancestors had often played a part; nineteenth-century female explorers were able to draw on a heritage of lands traversed and territories conquered. They took with them the confidence of a burgeoning age, as well as a certain amount of prejudice towards the cultures they would encounter.

One of Mary Kingsley's final lecture engagements before her premature death explored 'Imperialism in West Africa,' which was delivered at the grand and self-important Imperial Institute on Exhibition Row, South Kensington. Following her lecture, and with the subject of imperialism still on her mind, she expressed her beliefs in a thank-you letter to the Liberian magazine, *The New Africa* which had recently published a review of what was to be her final work, *West African Studies* – a letter that she wrote, rather tragically, on her final voyage out to the Cape. 'It seems to me a deplorable thing,' she wrote, 'that the present state of feeling between the two races should be so strained; and that unsatisfactory state, I cannot avoid thinking, arises largely from mutual misunderstanding. It does not seem to me to be unavoidable – a natural race hatred – but a thing removable by making the two people understand each other, and by avoiding rousing a hatred in either for the other by forcing them into interference with each other's institutions.'[34]

Kingsley's opinions were not well received, although her works were highly respected in the scholarly community. *The Times*' Colonial Editor, Flora Shaw, who herself had travelled widely in

South Africa, was purported to have refused to review Kingsley's work. As a staunch supporter of the British Empire, and an active promoter of emigration and the economic opportunities presented by colonization and expansionism, Shaw's jingoistic views did not sit well with Kingsley, who was reported to have commented that Shaw was 'imbued with the modern form of public imperialism. It is her religion'[35] and also hoped that Lord Lugard – a high-ranking colonial administrator who would become Shaw's husband in 1902 – would 'drown…in his own ink pot.'[36]

Kingsley's views were forward thinking for the day, but she was in the minority. She herself recognized this, stating quite plainly that 'I am not speaking for any one, only off on a little lone fight of my own against a state of affairs which I regard as a disgrace to my country.'[37] Where Kingsley understood the benefits of a tribal society, each with its own customs, rules and identity, the likes of Flora Shaw saw vast empty land, ready and ripe for industrious English settlers and economic enterprise. However, Kingsley was adamant that: 'No one else but the native can work its resources; you cannot live in it and colonise it. It would therefore be only an extremely interesting place for the zoologist, geologist, mineralogist etc. but a place of no good to anyone else in England…we have no desire to meddle with African domestic affairs, or take away their land,'[38] although she did concede that tolerating 'people who go in for cannibalism, slave-raiding, or human sacrifices'[39] would probably be a hard-sell to the rest of the British Empire. Her concern that the African people would become a 'landless people and an unhappy people' still resonates today as we seek to understand the legacy of Empire that we ourselves have inherited. For Kingsley, the solution was simple, and she made sure she expressed it in her final letter to the editor of *The New Africa*: 'I beg you, Sir, to do your best to prevent this fate falling on your noble race. I believe you can best do it by stating that there is an African law and an African culture; that the African has institutions and a state form of his own.'[40]

Of course, Kingsley's views were born out of the benefit of hindsight and experience. Whether she held the same opinions before she embarked on her travels it is impossible to know – after all, in her own words, she chose to travel to West Africa because that was where 'all authorities agreed' that the natives 'were at their wildest and worst,'[41] - but her 'skylarking' undoubtedly shone a light on European interference abroad, a narrative that remains troubling to this day.

As they prepared to depart most, if not all, nineteenth century explorers were not only carrying their luggage – they were also taking with them the cultural baggage of the age in which they lived, the impressions they had formed during their childhood and a worldview shaped by not only their friends and families, but in many cases their own imaginations.

The question was, would the rest of the world live up to their expectations?

Part 1

For the love of God

> 'Jane, you are docile, diligent, disinterested, faithful, constant, and courageous; very gentle, and very heroic; cease to mistrust yourself – I can trust you unreservedly. As a conductress of Indian schools, and a helper amongst Indian women, your assistance will be to me invaluable.'
>
> From *Jane Eyre,*
> Charlotte Bronte, 1847.

On the side of a bleak Tibetan hillside, Susie Carson Rijnhart gently lowered the medicine box into the ground, the lightness of the makeshift coffin betraying its tiny contents. As the last light of the day ebbed away into a shroud of darkness, Susie, her husband Petrus and their guide Rahim said their final goodbyes before covering the grave with earth and securing it with a boulder. The boulder wasn't to mark the final resting place of the poor lost soul within, rather it was to prevent wild animals from digging up the body and also to disguise the disturbed ground lest robbers believed there was loot buried there. In Susie's mind, the little grave did contain treasure, for it held the body of her beloved one year old son Charlie. 'With hands whose every touch throbbed with tenderness I robed baby in white Japanese flannel,' she later wrote, 'and laid him on his side in the coffin, where he looked so pure and calm as if he was in a sweet and restful sleep. In his hand was placed a little bunch of wild asters and blue poppies which Rahim gathered from the mountain side, and as the afternoon wore away he seemed to grow more beautiful and precious; but night was coming on and dangers threatened, and the last wrench must come.'[1]

Rijnhart's reflections of travelling as a missionary's wife in the final decade of the nineteenth-century contain some of the most heart-wrenching accounts of a married life dedicated to spreading the word of God. Her personal journey, both internally and externally, is one of unimaginable tragedy and truly horrific circumstances that would test even the hardiest of souls. Whether it was what Susie Carson Rijnhart expected when she married her missionary husband Petrus, it is difficult to tell. Certainly, the life of a missionary was well publicized and well understood by the late nineteenth-century, but Rijnhart's experience was particularly traumatic, and it is unsurprising that in her hour of need she supplicated God – after all, He was the reason that they were there, in the most inhospitable regions of Tibet. Susie tried her hardest to revive the child, 'feeling almost sure that nothing would avail but praying to Him who holds all life in His hands, to let us have our darling child. Did He not know how we loved him and could it be possible that the very joy of our life, the only human thing that made life and labor sweet amid the desolation and isolation of Tibet.'[2] With grim realization, the Rijnhart's 'tried to think of it euphemistically, we lifted our hearts in prayer, we tried to be submissive, but it was all so real – the one fact stared us in the face; it was written on the rocks; it reverberated through the mountain silence: Little Charlie was dead.'[3]

It wasn't the only hardship Rijnhart would face. In fact, the death of Charlie was only the beginning of her troubles, as we shall see.

Whilst the nineteenth-century is often recognized as the 'golden age' of the missionary, women have been travelling under the auspice of 'pilgrimages' since time immemorial. Roman citizen, Etheria, a noble woman from Gaul, journeyed to the Holy Land in AD 383 and was kind enough to leave us an account of her travels, which captures the gloomy, ritualistic essence of fourth-century Jerusalem, as well as providing a rudimentary holy tourist trail for others to follow. Etheria does her best to stick to the script of the devoted religious pilgrim, diligently recording in detail the many sacred rites, rituals and services conducted in the Holy Land, but she clearly cannot contain her excitement when confronted with the decadence of the liturgical objects:

> Now it would be superfluous to describe the adornment either of the church, or of the Anastasis, or of the Cross, or in Bethlehem on that day; you see there nothing but gold and gems and silk. For if you look at the veils, they are made wholly of silk striped with gold, and if you look at the curtains, they too are made wholly of silk striped with gold. The church vessels too, of every kind, gold and jeweled, are brought out on that day, and indeed, who could either reckon or describe the number and weight of the cereofala [candles] or of the cicindelae [oil lamps] or of the lucernae [lanterns] or of the various vessels?[4]

Her faith in God was unswerving, however, and like all good explorers she did her fair share of trekking and hiking in order to achieve her goals – which on one occasion meant ascending Sinai – but fortunately she had God on her side; 'Thus the toil was great, for I had to go up on foot, the ascent being impossible in the saddle, and yet I did not feel the toil, on the side of the ascent, I say, I did not feel the toil, because I realized that the desire which I had was being fulfilled at God's bidding.'[5]

One-thousand years later, Chaucer's gap-toothed Wife of Bath was doing much 'wandrynge by the weye,' (wandering by the way), having already made pilgrimages to Jerusalem (no less than three times), Rome, Bologna, Cologne and Santiago de Compostela. Her excursion to Canterbury must have seemed like a leisurely stroll compared to her continental adventures, which she undertook in her hat as large as a shield (she was 'wimpled-wel') and some tightly laced red stockings (her 'hosen weren of fyn scarlet reed') All of this meant, of course, that she probably had similar issues with clothing as her real-life Victorian successors - although there the similarity ends since it is unlikely they would have approved of her somewhat loose morals, particularly given their fervent Evangelism.

Come the nineteenth-century, Charlotte Bronte's depiction of St John Rivers in *Jane Eyre*, the pious but cold-hearted missionary-in-waiting embodied the values and principles of what was traditionally a male dominated profession. His rebuffed proposal to Jane, that she should accompany him to India as his wife because in his opinion: 'God and nature intended you for a missionary's wife. It is not personal, but mental endowments they have given you: you are formed for labour, not love,'[6] was typical of the early nineteenth-century, where many of the main missionary societies refused to employ single women to do God's work. Instead, they were expected to submit first to their future husband, and then accompany him on his religious mission. Jane Eyre subverts this stereotype by agreeing to accompany St John to India if she can 'go free' as his 'adopted sister' rather than as his 'helpmeet' and wife. Jane, very sensibly, does not agree to his terms – and thank goodness she didn't for if her body wasn't going to wither and die under the Indian sun, her soul most certainly would have if she were going to be eternally bound to the repressed and supercilious St John Rivers.

In the real world of the missionary, death was a very real possibility. Jane Eyre's assertion that, 'I should not live long in that climate…and to do as you wish me would, I begin to think, be almost equivalent to committing suicide,'[7] was depressingly close to the truth. Their commitment to delivering God's word could, quite literally, kill them. 'Death of a Lady Missionary,' the *Burton Chronicle* announced in September 1896. 'Miss Emma Entwhistle, a native of Berkenshaw, who was sent by the London Missionary Society to China, has died from small-pox, contracted from a person who attended her services at Sichaen. Miss Entwhistle intended coming to England next year to be married, afterwards returning to China.'[8] A similarly depressing missive was posted in the *Aberdeen Press and Journal* in 1894, although on this occasion, foul play was suspected. Scottish missionary, Miss Macarthur of the Southern Morocco Mission was thrown into the river Nifees while crossing a ford with a group of other missionaries, which included her sister. The party waited two hours, 'within which time the body did not rise to the surface.'[9] It later transpired that the unfortunate woman had been fished out – dead or alive, it was never established - downstream by a rogue band of Moors, who then hid her body and demanded a ransom of forty dollars for her release; 'The money was paid, but when restored [to the party] she was dead,'[10] reported the *Manchester Courier*.

Bandits and robbers were often the least of a lady missionary's worries. The government of the 'host' country could prove to be just as unwelcoming and, on some occasions, extremely hostile. In 1889, the *Manchester Courier and Lancashire General Advertiser* reported that the Korean government had made a policy of arresting and charging Christian missionaries if they attempted to distribute Christian propaganda to its citizens. A Mrs Heron, of the American mission, was purported to be the first victim of the policy when she was, 'arrested on a charge of making converts to Christianity, and, after what is described as a trial, was actually sentenced to death.'[11] Fortunately for Mrs Heron, the Americans had a diplomatic presence in Korea and no doubt after much shuffling of official papers and verbal haggling, the condemned lady was allowed to go free. If not, the newspaper predicted 'stern reprisals.'[12]

It is all too easy for us twenty-first century beings to scoff at these early attempts at do-gooding, but it is disingenuous – and a little hasty - to tar all missionaries with the same meddling brush. After all, their not insignificant efforts came from a genuine belief that the word of God would improve the lives and moral fibre of the local 'heathens' they wished to convert. While their top-hatted industrialist fathers and brothers were busy building an empire at home, women were charged with carrying a message of hope to its furthest corners. It was a time of progress, resourcefulness and vigour, and God did not look kindly on the idle.

It was also one of the few career options available to an unmarried nineteenth-century woman, providing not only an alternative to marriage or the dreaded spinsterhood, but also an acceptable excuse to board a ship and see something of the world. The image of the nineteenth-century Christian do-gooder as either a stout, sanctimonious, matronly sort, or a meek, slightly feeble looking plain Jane à la the eponymous Miss Eyre, spring to mind. This is wholly unfair, of course, and as social historian Jane Robinson is keen to point out in her definitive guide to women travellers, their accounts of their journeys 'are amongst the most uplifting and exciting of all travel books...the description of it [their journey] is rarely sanctimonious; despite the physical and emotional stresses involved and, often, the apparent lack of success in terms of 'converts,' there is a spirit of confidence and affectionate humour...'[13] Nonetheless, being seen to be meek and humble was to their benefit since it was hoped that as members of the fairer sex, they would present less of a threat to the local people - although for our American missionary on Korean death row, Mrs Heron, this clearly wouldn't be the case.

The missionary also faced her fair share of derision from home quarters. In April 1892, the *Toronto Daily Mail* reported 'A Pertinent Question.' During a recruitment talk given to a lady's club by a female missionary recently returned from South Africa, the native people were described as 'gentle, trustful folk, honest, affectionate, and moral, not wanting, in fact, in the simple elements of character.'[14] Following her earnest appeal for help rescuing the 'moral folk,' 'from heathenism and savagery,' a member of the audience quite rightly asked 'Why should we take these people out of such Arcadian simplicity? Is it to give them corsets and the catechism?'[15] Quite.

Those who decided to stay at home and give marriage or spinsterhood a try could be 'consoled' by the knowledge that their life choices were viewed by some in the religious community to be just as hard, if not harder than those of their religious sisters abroad. In 1896, *The Young Woman* published a sermon (of sorts) from the Rev. A. R. Buckland – a preacher at the Foundling Hospital. His advice, under the title 'Women and the Leaven,' was that service to God, more than anything else, began at home. According to the Rev., being at the centre of the domestic sphere presented 'great opportunities' and 'very special openings for doing their Master's service.'[16] He then proceeded to list the opportunities, which included 'sickness, family anxieties' and 'personal disappointments'[17] – all of which, in his opinion, would enable a woman of faith to serve her Lord. And if her faith started to flag, she could be reassured that her sex 'are so long-suffering and show such vast staying powers that it does not seem cruel to remind them of the need of a patient continuance in well-doing, however uncongenial the atmosphere in which they must show it.'[18] The Reverend very generously suggested that those who did go abroad would find their situation easier because 'the limitations provided by separation from home and kindred, by contact with the heathen, and by the settled round of exacting toil; the stimulus of carefully selected companionship, and the knowledge that the worker is upheld by the prayers of many at home – these things all help to lighten the stern toil of the woman missionary.'[19] Whether those in the field agreed with his sentiments is, arguably, doubtful.

So, with such a damning record of deaths, murders, life-threatening illnesses and general mishaps – all of which make for a particularly off-putting job description for a female missionary – and an alternative option of finding faith in the drudgery of everyday life should they choose to remain at home, it is intriguing to examine their motives and consequently, how the strength of their faith showed them the way in some of the most inhospitable and fundamentally hostile regions of the world. With God on their side, they were certain to achieve their holy goals, whatever hardships they had to endure along the way. Or were they?

Let us return to the tragic story of Susie Carson Rijnhart, who was born in Ontario in 1868, and at the tender age of twenty graduated from Trinity College in Toronto as a medical doctor. Her faith was established in childhood, and she was an active member of the Methodist churches in both Chatham and Strathroy in her native Ontario. Certainly, by the time she embarked on her work in Tibet, the shared faith of Susie and her Protestant Dutch husband, Petrus had led them to their calling and would, they trusted, enable them to survive their first mission together in 1895;

> We went forth, however, with a conviction which amounted to absolute trust that God would fulfil His promise to those who 'seek first the Kingdom,' and continue to supply us with all things necessary for carrying on the work to which He had called us. From the outset we felt that we were 'thrust forth 'specially for pioneer work, and although anticipating difficulties and sacrifices we were filled with joy at the prospect of sowing precious seed on new ground.[20]

The couple reached the village of Lusar, near the Kumbum monastery on the Sino-Tibetan border and, having set themselves up with a temporary home, opened a dispensary – a decision which proved itself provident in terms of helping the sick and administering to their souls at the same time. 'As soon as the people found out that we were prepared to treat their ailments and dispense medicines they came to us quite freely,' wrote Rijnhart later. 'The Chinese were the first to approach us, but soon the Tibetans came, even the lamas, and it was not long before we had as much medical and resultant guest-room work as we could attend to.'[21] To the dedicated couple, every patient presented an ideal opportunity to preach the word of God, and as Rijnhart remarked later: 'it is impossible to get a crowd of Tibetans to listen to a discourse, so our evangelistic work consisted chiefly in conversing upon Christianity with the people who came to see us, and from the very beginning we were able to interest them in the teachings of the New Testament.'[22] For a country already overloaded with religions, introducing Christianity was always going to be a bit tricky.

Apart from a Muslim revolt in 1895/6 it was so far, so good for the devout duo and their son, Charlie, who was born in 1897 and 'brought with him a budget of love and a stock of sunshine,' which only added to their joy.

The Rijnhart's calamities began in 1898. Their goal was to reach the sacred and forbidden city of Lhasa, a feat that hadn't been accomplished by a Westerner since the Jesuit missionaries Huc and Gabet visited in 1846. They *could* have remained on the fringes of the sealed land but the pull of the mystical capital and all that it embodied was overwhelming. The Rijnharts drew strength from the challenge, rather than abandoning their hopes on the side of a Tibetan hill – a decision, which with the benefit of hindsight was, at best, brave, at worst, foolhardy. It became something of a crusade, for they would surely succeed where others would fail – they just needed to conquer the terrain and then the Tibetan officials, in the hope that events would eventually 'lead to the downfall of the barriers that too long have kept a people in darkness, and bid defiance to the march of Christian civilization.'[23] And it was to be the Rijnharts leading that march, under God's command to 'Preach the gospel to every creature.' Susie later wrote that:

> We asked that we might be divinely guided at every step and that the means might be provided for the journey. Our prayers were answered and, although we knew not what the results would be, we rejoiced exceedingly that we were counted worthy to traverse for the first time in the name of Christ whole districts in which His name had never been heard.[24]

Despite God's divine intervention, and much travelling experience between them to draw upon, the odds were stacked against them. Susie herself stated pragmatically that 'there was absolutely nothing inviting in such an undertaking.' Lhasa was approximately eight-hundred miles as the crow flies away, but the route lay across a series of mountain ranges and passes of up to 5000ft. Nevertheless, they set off on 20 May 1898 with their son Charlie, now ten months old and kitted out in a 'little fur *ja-ja* [sleeveless jacket]', three guides, several horses, enough food for a year, and 500 New Testaments translated into Tibetan.

Three months into the journey and two-hundred miles from their destination two of their guides deserted them whilst they were crossing the Kuenlen Mountains. Soon afterwards, five of their horses were stolen and then, in Susie's words, 'the darkest day in our history arose'[25] when baby Charlie died. Susie's distress is palpable in her heartbreaking account of his burial as 'the cold earth of Tibet, the great forbidden land, closed over the body of the first Christian child committed to its bosom – little Charles Carson Rijnhart, aged one year, one month and twenty-two days.'[26]

This would have been trial enough, but fate (or God) had more tests in store for the unfortunate couple. Several days later, and no doubt labouring under an overwhelming sense of grief, their mission to Lhasa ended abruptly when officials – who to their credit, could have killed the couple on the spot – pitched up next to the Rijnhart's camp and erected a tent before ordering the Rijnharts to turn around and make their way back to China. This devastating blow was delivered in the sumptuous surroundings of the Tibetan's white tent, in which the *ponbo ch'enpo* (Tibetan chief) sat on a dais, surrounded by richly decorated Turkish rugs and mats. Having exchanged pleasantries over that unifying global beverage – tea (in this case directly from the *ponbo's* own teapot) – the chief informed them that should he allow them to continue their journey to Lhasa, he would be beheaded. To this, Mr Rijnhart responded that 'he would prefer being beheaded to returning by the route over which we had come.'[27]

Touché.

The two camps – the Rijnharts and the Tibetan officials – then proceeded to have the politest standoff in recorded history, in which both sides exchanged gifts (bags of rice, tea, butter and a 'big fat sheep'[28] for the Rijnharts; a bound copy of the Gospel's and a scarf for the *ponbo*). The turning point came when the Rijnharts were invited to the *ponbo's* tent and were served tea – not from the *ponbo's* personal teapot, but from a different one. This small act signified that their days were numbered, but they did at least manage to negotiate new horses and three guides for what promised to be a terminally dull and frustrating return journey to China.

After trekking for several days through a blizzard and heavy snow to reach the destination agreed with the *ponbo* – the monastery of Tashi Gompa - they were attacked by Tibetan robbers, who shot at and threw boulders at them from the hills above. The bullets narrowly missed Mr Rijnhart but injured a guide and killed all but one of their horses. The guides then deserted them, no doubt out of fear for their own lives, meaning the Rijnharts had no way of knowing how to get to Tashi Gompa. They stumbled on blindly, camping where they could and managing to survive the inhospitable climate and rugged terrain before tragedy struck again.

Having, with some relief, spotted some tents on the other side of a ford, the Rijnharts decided to pitch their camp for the night. The following morning, Petrus attempted a crossing with the aim of hiring some more animals and getting directions to Tashi Gompa. He left Susie with the 'big revolver,' telling her not to be afraid and that he would return before nightfall, if possible – but if not, he would call out when near so as not to frighten her. In Susie's own words:

> When a few steps away he turned to wave his hand and said 'ta-ta'…Wading half across, he put out his arms to make the first stroke, but suddenly turned around and walked back again to the bank where he had first entered the water. Shouting something up to me which I did not hear on account of the rushing river, he walked up-stream in the opposite direction to the tents he had set out for. Then he followed a little path around the rocks that had obstructed our way the day before, until out of sight, and I never saw him again.[29]

It is impossible to speculate on just how painful an ordeal it would have been for Susie Rijnhart during those hours and days as she waited for Petrus' return. She managed to take comfort in imagining various scenarios – perhaps he was merely enjoying a cup of hot tea and the hospitality of some friendly Tibetans, or perhaps he was still bargaining for animals – but inevitably those thoughts turned darker, until she had no option but to face the reality of her dire situation.

Now completely alone in a hostile region, already burdened with the grief of losing her child, Susie somehow clung to her faith. As she watched the sun go down and the shadows grow long, she 'prayed for strength to be quiet, for God to give me freedom from anxiety.'[30]

Petrus was gone - missing, assumed dead. Susie then sat on that bleak hillside for several days, alone with her thoughts. Her heart-wrenching but lyrical description of those moments, which were written after the mission and thus with the benefit of time, distance and reflection, belie the truly horrific circumstances that she found herself in. She was a recently bereaved female, lost and alone, in a hostile country in which thieves and murderers roamed at will. French explorer Jules-Léon Dutreuil de Rhins had been murdered four years earlier near Xining and his body carelessly dumped in a river, and Annie Taylor, a contemporary Tibetan traveller of the Rijnharts, had only narrowly avoided being stoned as a witch by the people of Tashi Gomba. There was no chapter in Campbell Davidson's *Hints to Lady Travellers* that could be of any use to Susie Rijnhart during those desperate hours. She wrote:

> Dusk settled into darkness, and a desolate solitude reigned over hill and valley, almost chilling me to the heart as I sat alone in the stillness of that oriental night, broken by no sound of human voice, with no sympathy of friends to fall back upon.[31]

She may not have had friends to fall back on, but she always had God. For what, or who else, could steer her through the darkness and back to the light.

And much to Susie's relief, God did eventually deliver her to safety. Note - *eventually*. Frankly, it was the least he could do given the circumstances. Her terrifying ordeal didn't quite end with Petrus' disappearance, however. First, she had to contend with a succession of awful guides – some of which had designs on her meagre possessions, her body and her life, and second, she had to watch the last of her husband's keepsakes, his diary and Bible, be condemned to a 'miry stream' lest they be discovered and Susie's identity as a foreigner exposed.

On 26 November 1898, two months after Petrus' disappearance, Dr Susie Rijnhart stumbled into the China Inland Mission in Ta-chien-lu, penniless, in rags and with frost-bitten feet. Her ordeal was over. She stayed in China just long enough to hear the outcome of the investigation into her husband's death, which unsurprisingly yielded no definitive answers other than what Susie already knew, and by 1900 she was back in Canada.

With her faith still intact and still willing to fight the good fight of the Lord ('Christ also has his soldiers who are willing to die for his cause if need be, in the belief that his cause is the sublimest among men'), Susie returned to China in 1902 and by 1905 had married another missionary, James Moyes. However, her health was, understandably (given her history), delicate and led to the couple's premature return to Canada in 1907. In January 1908, Susie gave birth to a baby – another son. It was the happy conclusion that everyone was hoping for.

Except it wasn't.

Susie died just three weeks after giving birth in February 1908. For her to have survived such a horrific experience, but then suffer such a clichèd fate proves that God really does move in the most mysterious of ways.

Annie Boyle Hore also found out the hard way that missionary work and motherhood seldom make an agreeable pairing. Although Susie Rijnhart had delighted in having her son Charlie as a travelling companion until his sudden death, it is safe to say Mrs Hore's experience was less cheerful, although everyone did at least live to tell the tale. Annie's account of her journey to Ujiji, which she published under the charmingly titled *To Lake Tanganyika in a Bath Chair,* relates the slightly unusual circumstances under which she found herself travelling into the African interior.

At the behest of her husband and his employers, the London Missionary Society (LMS), she was to be a science experiment. The Hores were not missionaries themselves – Annie's husband Edward was, in fact, a scientist – but he was charged by the LMS with assessing the viability of prospective routes for missionary groups, which was almost certainly a nice way of saying that whether the routes were feasible or not depended on whether they all made it back from the crocodile-infested swamps in one piece. This was the 'knotty question'[32] that the Hores were required to answer.

Annie was a necessary appendage to this slightly barmy scheme since Edward's remit included finding out if women could undertake the routes. Three-month-old baby Jack was there by default, but his survival would be a useful barometer of just how feasible the journey was. Edward was convinced that wheels were the answer and in fact, his zeal for the mission was such that he declared at a public meeting that 'if he could succeed in getting no other vehicle, he would at least take his wife to Ujiji in a wheelbarrow' – a prospect that no doubt filled Annie with glee. They eventually settled on a wicker bath chair, and baby Jack was relegated to the wheelbarrow instead.

The bath chair was kitted out with sturdy wheels, padding and a hood and Annie later wrote that 'of all the means of conveyance I had tried, the bath chair was by far the most comfortable, except when in motion…'[33] which was a bit of a problem, given the aim of the experiment.

Annie's epic bath chair trek had several false starts, but she eventually succeeded in being carried all the way in a rudimentary sedan chair, which was uncomfortable enough even without the fact that baby Jack 'would stop nowhere but on my lap…and all the detail of nursing him had to be effected in the narrow compass of the bath-chair.'[34] He obviously wasn't keen on the wheelbarrow idea either.

In between dealing with the needs of a fractious child in the searing heat of the African interior, Annie could at least enjoy the scenery - except for the 'curious dark objects lying on, or beside the path' that turned out to be the 'dead bodies of helpless laggards from the various hungry caravans that had passed that way. The heat and drought had been so great, that these bodies were perfectly hardened and preserved.'[35] The poignancy and significance of the 'dreary scenery' was not lost on Annie. Her pragmatic approach to her experience meant that her world view was wide enough to appreciate the darker forces at work in Africa at that time.

> It was a terrible sight, which suggested horrors worse than mere death in connection with the diabolical system of man-hunting, and the driving of the victims in herds, on the speculation of a good percentage surviving to arrive at market. I do not mean to describe the horrors of the slave traffic, for I fear I cannot bring about the effects which so many more eloquent witnesses have failed to produce, but I must discharge my conscience of this duty to solemnly remind and warn whoever in Christian and civilized lands may read this book, that inner Africa is as much as ever given over to all the horrors of the slave traffic, so often and so ably described.[36]

Despite her efforts – the journey she recounts was the third and final attempt to reach Ujiji – she received a mixed response. The *Pall Mall Gazette* applauded her 'courageous act' and 'successful journey,' recognizing the 'very useful work that a woman, and a woman only, can do in missionary enterprise.'[37] And whilst the *Gazette* was less than complimentary of her writing style – 'Mrs Hore is no fine writer' - they at least recognized the merits of her honest appraisal of African travel and 'its hardships, its succession of wearying troubles and disappointments, and its rare enjoyment.'[38]

Not so for *The Field, The Country Gentleman's Newspaper*, who were a touch more scathing. 'No part of the journey was performed on wheels,'[39] they declared, before likening Annie's chosen method of transport to that of an invalid – the inference being that she herself was being a bit feeble. 'Indeed, the journey was performed in the manner usually adopted in the country by sick persons, or those unable to bear the fatigues of the journey on foot.'[40]

The misogyny did not end there, either.

> The journey was, as compared with many other African journeys, uneventful, and no expense appears to have been spared to procure all the comforts the authoress required. One of the saddest features in the narrative is the manner which poor little Jack, after having had African fever on each of his two former journeys into the interior, was taken for the third time into that pestilent country, only to get the fever again, and in that miserable state to be carried, under the burning sun, or through salacious swamps, to Ujiji. What good this poor child could do to the missionary cause, or why he should have been taken [on] that weary journey, is best known to his parents.[41]

In the eyes of the press, not only was Annie lazy, but she was also an irresponsible parent. The 'uneventful' journey, which included the threat of hyena and crocodile attack, a gun fight, an enormous flood, famine and severe illness was eventually concluded to everyone's satisfaction – especially little Jack's, who now at two years old was thrilled by the heavy rain and floods because it meant his wish to ride in a canoe would be granted. 'We are in the baby boat, Ma,' he exclaimed with glee. But most importantly for the Hores, the experiment had been a success – despite it being considered by Annie's friends back home as 'utterly ridiculous.'[42]

Annie's take on missionary work was a thoroughly modern one. Once settled in their new abode (a mud hut of dubious construction) in Kavala, she soon realized that the well-meaning advice she had received back in England of setting an example of household industry to the local women would require her to model that behaviour herself.

Instead, she decided she would 'lose influence and respect entirely by doing such things myself, and that the drudgery and overwork of women here is already such that a better example would be to set to the natives if I could persuade my husband to do the laundry himself.'[43] Annie was sympathetic and perceptive. Her matter-of-fact approach, which left no room for complaining or self-pity, lends credibility to her account and the judgements she makes on her surroundings. Annie's mission in life was to serve her husband, rather than God – and serve him she did, ungrudgingly and with barely a groan.

Neither the *Pall Mall Gazette* or *The Field, The Country Gentleman's Newspaper* remarked on Annie's revelations that the slave trade in Africa was still 'festering' – she had seen them herself, 'absorbed along the road, where anyone who looks may see their carcasses; and otherwise amongst well-to-do tribes, where every family has one or two, the result of murder and pillage at some distant place.' Annie was also insightful enough to recognize that the 'civilizing' influence of western missionaries was also nurturing another destructive force – consumerism. Wherever there was a missionary station, there was also a gathering of natives 'both in hope of more peaceful times and for the benefits to be derived from the presence of the missionaries.' Not much escaped her notice and her sensitivity to the nuances of the missionary-native relationship lends itself to the authenticity of her voice.

With the experiment over, Annie closes her account with a final homage to the vessel that carried her 830 miles in ninety days: 'The bath chair is put away in the store, and my story must end.' And with that, like many of her adventuring comrades, she disappears from view.

Where Annie Hore was decidedly lax in her missionary endeavours, Annie Royle Taylor was anything but. The epitome of iron-willed evangelicalism, Miss Taylor was committed to achieving salvation for the Tibetan heathens from the age of thirteen. While other girls of her class were mastering their pony-riding skills or sewing quietly in a corner, young Annie was masterminding how to spread the Gospel to the pagans living in the mystical heart of Tibet. In her own words, she was a 'troublesome child and very full of mischief,'[44] and no doubt caused her nurses and governesses a great deal of angst. This was compounded by the family doctor, who made the mistake of advising her parents that her medical condition – valvular heart disease – would kill her in childhood, and she shouldn't be troubled with studying: 'So for some years I did much as I liked,' she wrote. The devil had made work for idle hands, but that was set to change.

Annie's father, John Taylor, had slightly different expectations for his wilful daughter – after all, he was the director of a successful fleet of sailing ships and probably the proud owner of several stove pipe hats. That his daughter wished to sacrifice herself to the Lord's work in the most inhospitable country in the world, rather than enjoy the bourgeois comfort of a well-to-do mid-Victorian household became the source of twelve years of father-daughter feuding, before Annie finally got her way.

In the meantime, Annie earned her missionary stripes tending to the sick in the slums of Brighton and London, and studied medicine at the London Hospital in Whitechapel and Queen Charlotte's in Marylebone, from which she graduated in 1884. When her father threatened to stop her allowance, Annie deployed the ultimate act of adolescent defiance by selling her jewellery and moving into her own lodgings. It was acceptable to break the Second Commandment to 'Honour thy mother and thy father,' if it meant she could get on a boat to Shanghai and join the China Inland Mission. She could always make up for it later – which of course, she did, in the Godliest way.

With her father's will broken, along with the Commandment, Annie finally set sail in September 1884. She was quite the spectacle, being an unmarried female missionary, and such was John Taylor's pessimistic conviction that Annie's sojourn would be short-lived, that he offered to buy her a return fare. Annie did eventually return – but not for twenty years. Her first posting was to Chinkiang, a port near the mouth of the Yangtze where she mastered Chinese, and joined in the fervent gatherings at the Mission. These 'gatherings' involved the hearty singing of hymns, lecturing on the scriptures and bible study, and often lasted the entire day 'until the lady missionaries were faint and hoarse.'[45]

One missionary, a Mrs Nicoll from Chungking, was so overcome that she did actually collapse with exhaustion and had to be fanned back to life by a group of Chinese women.[46] The Mission's motto was *Jehovah Jireh* – 'The Lord will provide' – and it was with this belief that Annie was sent out to evangelize the pagan masses. It was a motto she would fall back on regularly as she navigated her way through her religious mission and the inhospitable landscapes of East Asia.

Annie had big plans. By 1886 she had been posted to Lanchow, the capital of Kansu province near the Tibetan border, where she took on the sole responsibility of saving the souls of its female inhabitants. 'Lanchau is a very wicked city,' she wrote, 'the marriage tie is not looked on as sacred.' Witnessing first-hand the exploitation of its women-folk, she asked, 'Who will come and raise them? They are going down, down into hell, and there is no one but me in this great city to witness for Jesus amongst them.'[47]

In 1886 she visited the Great Lama Monastery of Kumbum, but her true purpose lay beyond the frontier. Despite the dangers, she wanted to penetrate the dark unevangelized heart of Tibet and make an attempt on Lhasa. This, she reflected, would pose no problems: 'We have received no orders from the Lord that are impossible to be carried out.'[48] 'I went,' she wrote, 'in simple faith, believing that the Lord had called me. I knew that the difficulties were great, and that enemies would be numerous, but I trusted God to take care of me, just as He protected David from the hands of Saul.'[49]

She would need God's protection from the outset, and luckily, she got it. She pressed forward into Sikkim, where the Tibetan frontier authorities regarded her with suspicion, as did the natives who regularly asked her what they should do with her body if she died. This was important information, because they had already told her they were planning on 'praying her dead,' as apparently was customary, 'and to this they resorted, taking care to help their prayers in a very effective manner.'[50]

When the prayers didn't work, they resorted to a more traditional method. The local chieftain's wife invited Annie to have a meal with them, but as she was eating, Annie's suspicions were aroused by the eggs and some conversation she overheard between the assembled company. 'Sure enough, after she had partaken, she became ill, with all the symptoms of aconite poisoning,'[51] *The Christian* reported in August 1893. She did, at least, learn Tibetan when she wasn't trying to avoid her own death, and managed to make a friend of a Tibetan called Pontso, so it wasn't all bad. Pontso was also a native of Lhasa, so in missionary terms, it must have been like hitting the jackpot. Unsurprisingly, Annie was quick to convert him into a 'follower of Jesus.' After ten months in Sikkim, Annie and Pontso sailed to Tauchau, a city on the Tibetan border, and after a further year of preaching to some 'intelligent lamas, free from the grosser superstitions,' they decided to make their attempt on Lhasa. This they would do with the help of a Chinese Mohammedan called Noga who spun Annie what, on the face of it, appears to have been a bit of a tall tale. Allegedly, Noga had promised his mother-in-law that he would bring her daughter, his wife, back to Lhasa, where she had resided before their marriage but, alas, he had no money, and so he consented to conduct Annie to Lhasa, provided she found them horses and paid for the privilege. At sunrise on 2 September 1892, the group which also included some hired servants, stole across the frontier into the same bleak landscape that would claim Susie Rijnhart's baby six years later.

Unfortunately for Annie, Noga turned out to be an utter blaggard: 'a faithful and false guide,' who 'proved to be a great rascal, whose only object in taking Miss Taylor into Tibet appeared to be to rob and murder her.'[52] Luckily, Noga only managed the first of those offences and Annie escaped with her life – maybe she was protected by the Lord after all. God's navigational skills also came in handy when Annie got lost for three days in the mountains ('this had been God's method of sheltering them from a deliberate attempt at murder'[53]) and even though the horses and yaks had taken to eating clothing as a substitute for grass, with one of them growing so tired it literally lay down and died beneath her ('it looks as though I could not go on without a good horse. God will provide me with one'[54]), her faith was never in question. She was remarkably dismissive of the Tibetan's beliefs – on the one hand recognizing that 'The Tibetan's are a religious people,' but on the other deriding their rituals: 'I was shocked to see men and women near Ta'ri'si, prostrating themselves the whole length of the road…Poor things, they know no better; no one has ever told them of Jesus…' Saving their souls was her only goal and she had no intention of being distracted by attempting to understand their own sacred rites and cultural ceremonies.

Annie endured endless attacks by bandits, and it was so cold that touching a knife made her skin come off but, 'The Lord is good,' she confided in her diary. 'I will not fear what man shall do unto me.' So that was that. They waited three days for a river level to fall so they could cross it ('We are waiting until tomorrow to see if it will be lower in the morning. The Lord can do this for me.'[55]) If he could do it for the children of Israel in the Red Sea, he could do it for Annie.

One of her servants died – his end hastened by the rascally Noga, who insisted on bathing the poor man in sub-zero temperatures before he died, as was customary for the Mohammedans. Then they had to work out how to bury him in the frozen ground. They finally settled on 'a piece of swampy ground,'[56] using a few sods to cover him up 'but not before the wolves were all howling around.'[57] Still, she found a way to enjoy 'A nice Christmas day' on Jesus's birthday and boiled a Christmas pudding in his honour.

To cap it all, Annie was not only robbed by Noga, but he also betrayed her to the officials in Lhasa, and Annie, alongside Pontso and another servant named Penting, was remanded in custody within three days reach of the sacred capital. Annie remained utterly unfazed, remarking that the chief who held her in custody was 'quite a dandy in his way, does his hair with a fringe in the front and a fantastic plait at the back,'[58] and was pleased to be offered mutton and some cheese to eat, which must have been a blessed relief after the barley flour with mushrooms they had been eating up until that point.

But Annie knew she was defeated – this time at least. She was sent packing with a contingent of fresh horses, provisions, a tent, and an escort of ten soldiers to make sure that she left Tibet. And so, on 18 January 1893, Annie Taylor, missionary extraordinaire, turned back the way she had come, finally entering the Chinese town of Jyekundo on 21 February and from there, journeying on to Ta-chien-lu, which lay just over the Chinese frontier. She arrived there on 13 April, after a journey of 1300 miles that had taken seven months and ten days to accomplish. After a year's sojourn in England, during which Pontso accompanied her, she returned to Tibet under her own mission, catchily named the 'Tibetan Pioneer Band.' She eventually settled in Yatung in a ramshackle hut she named 'Lhasa Villa' where her sister, Susette Taylor was reunited with her in 1903. 'I suddenly become aware of an advancing figure that exactly suits the surroundings,' Susette wrote of their meeting. 'A brick-red gown of native cloth, with a glimpse of fawn silk at neck and wrists pouched up above the girdle, thus displaying blue cloth trousers tucked into fur boots the shape of night-socks, drapes a small person with a merry face much too fair for a native, and topped by a yellow peaked cap. It is my sister! She greets me affectionately, asks if I have had a pleasant journey – much as if she were meeting me at Victoria Station – and takes me to the Mission House…'[59]

Annie was by now such a familiar figure in the town that she was known to all as Yatung *anni*, the Tibetan word for aunt and a handy pun on her own name. Sometime before 1909 she returned to England in poor health but like so many of her kind, she ended her life in obscurity, spending her final years at a private asylum in Fulham, London. She died on 9 September 1922 and was buried four days later at West Norwood Cemetery.

No doubt she regretted her strange and harrowing adventures not one bit. And she'd also proved that women were just as well suited to missionary life abroad as men – if of course they were that way inclined. After all, they also had the option of staying at home and saving the souls of some English heathens instead – because by the end of the nineteenth century, there were just as many unholy places lurking in Britain as there were abroad, most of which didn't require the presence of yaks or bath chairs to reach. It was the challenge that drew them out to the faithless – not just the challenge of conversion but the challenge of conquering the landscape. Religion might have been their primary motivation, but a desire to travel and explore was just as important to the Victorian female missionary. Was there a hint of wanderlust pulsing through the female missionary's veins? One can assume so - it was just hidden under the respectable cloak of faith and moral purpose – whether the female missionary was prepared to admit to it or not. Regardless of what drove her onward, she was undoubtedly a force to be reckoned with and just as likely to go beyond the garden gate as her more adventure-loving contemporaries. As *The Housewife* said of Annie Taylor in August 1893, 'Not only has this intrepid woman traveller thus travelled thousands of miles of country where no European foot has ever before trod, often amid races devoted wholly to plunder and fighting, but she has done so without a single companion of her own race. Miss Taylor certainly does not look a typical traveller.'

There was nothing typical about Annie, so there is no doubt she fitted the explorer mould, whether she was aware of it or not.

Part 2

For the love of a good (or bad) man

'I was quite infatuated by this man, and I might never have such a lover as he was again, but I was obdurate. 'Well, you are a woman of spirit. Come with me again this evening, only for a little while,' he said. He asked if I was annoyed because he had come to my room last night, and I said I was. 'But you said yourself that I could come,' he replied.'

From *Love Among the Butterflies: The Travels and Adventures of a Victorian Lady,*

Margaret Fountaine, 1980.

In November 1940, the Castle Museum in Norwich took delivery of a padlocked black japanned box, which came with the strictest of instructions that it was not to be opened until 15 April 1978. It was accompanied by ten mahogany display cases containing 22,000 butterflies – the lifework of its curator, Margaret Fountaine - which would be named, according to the bequest, the Fountaine-Neimy collection.

The legacy was put into storage, where it sat patiently for over thirty-seven years before being duly retrieved on the appointed date, as per the conditions of Miss Fountaine's Will. Fountaine was known to be a little eccentric and the museum had low expectations of the secret contents – an additional species of butterfly, perhaps? Or was it possible that the box contained the remnants of her entomological notes? Either way, all involved were, at first glance, underwhelmed by what predictably turned out to be just another box of papers. The *Daily Mirror* picked up the story, telling it in their usual bawdy style,

> An old tin box was opened yesterday…and out spilled the secret loves of Madam Butterfly. The box contained the diaries of celebrated Victorian butterfly collector Miss Margaret Fountaine. Miss Margaret, a rector's daughter, scorned danger as she trekked through the world's wildernesses with her butterfly net. But flitting through the pages of the diaries is a colourful collection of MEN who set her heart a-flutter. Men like Bellacosta the Corsican bandit and Septimus the passionate Irishman. Sadly, the diaries outline a life of disappointed love.[1]

When Margaret was found dying, butterfly net in hand, at the side of a road near Mount St Benedict on the island of Trinidad in 1940, she had already set the wheels in motion for her unusual bequest – ordering that the secret box and its even more secret contents be opened one hundred years after the first diary entry on 15 April 1878.

The diaries comprised twelve thick, ledger-like volumes which contained more than a million words in Margaret's hand, along with photos, drawings, pressed flowers and postcards. Words such as the quote at the beginning of this chapter, which would eventually be published posthumously as *Love Among the Butterflies: The Travels and Adventures of a Victorian Lady.* Words that, given the time of their writing, were quite remarkable for their candour.

Margaret Elizabeth Fountaine was born in 1862 in a period of great scientific progression and a blossoming awareness of evolutionary theory. Her father's position, as the Reverend John Fountaine of South Acre parish in Norfolk, provided Margaret with the good fortune of being both respectable and of moderate means. Her diaries, at a glance, read much like any other narrative of Victorian girlhood - filled with parties, family gatherings of numerous aunts, uncles and cousins, church-going and sibling squabbles. Yet Fountaine's early descriptions of domesticity contradict the spirit of adventure that would eventually lead her to traverse the globe for her two great lifelong passions: lepidoptery and love.

When Margaret was fifteen her father died, and her first diary entry on 15 April 1878 documents the family's move from the old rectory into their new house, Eaton Grange in Norwich. Still very much a child, Margaret comments on being told off by her mother for making too much noise on the uncarpeted floor, and how they 'all had tea in the schoolroom, a nice light airy room. I like my bedroom, which has just been freshly papered. About half past eight we went to bed, and here ends the account of how I spent the 15th of April 1878, so good-bye for three hundred and sixty-four days.'[2]

A year later and teenage angst had kicked in for poor Margaret,

> Oh! How differently I feel now to what I did when I wrote that…The very streets of Norwich have become sacred to me, because in that year I have learnt a new feeling, a new love so great that it cannot confine itself to the one object but spreads over everything that is in any way connected with it. I love the long straight road because it is there that we generally see 'him'…I love the little church because 'he' has been there. Oh, we never knew at South Acre what real happiness was.[3]

Fortunately for Margaret, the widowed Mrs Fountaine – 'Mamma' to Margaret - had her hands too full with Margaret's seven siblings to notice that her teenage daughter was in danger of making an Unsuitable Attachment to a certain Mr Woodrow, and so Margaret was able to indulge her crush for a full year before she was forced to admit to her mother that she 'loved Woodrow far better than anyone else in the world,' even though she 'had never so much as exchanged a single word with him.'[4] Mamma wasted no time in dousing the flames of eternal love by informing her lovesick daughter that even if he 'had a fortune of £20,000 a year, she would have thought it hardly possible for any of us to fall in love with him, and had always congratulated herself on his being quite safe.'[5] Which is the Victorian equivalent of a reality check.

A year later in 1881 and as is customary with teenage crushes, Mr Woodrow has been entirely forgotten about – which is handy because Margaret confides that he has gone to Australia to make his way in the world. 'I really believe I should have died of misery at such a separation, but happily I had ceased to take any interest in him,' she writes pragmatically.

Except love wasn't quite finished with Margaret. If he had a pulse then she was certain to fall for him; if it wasn't the son of a family friend, then it was the local (married) curate; until 'another image haunted my imagination,' that of Irish chorister, Septimus Hewson. For Margaret, Hewson was the real deal and she contrived to bump into Mr Hewson at every available opportunity, even if it involved hanging around street corners or skulking around the cloisters of Norwich Cathedral with her sketchbook for cover. She did eventually pluck up the courage to speak to Septimus and was set all a quiver when he 'held my hand a second or two and squeezed so tightly I could positively feel the strength of his muscles...'[6] It was all most unseemly for a genteel Victorian lady.

Mamma wisely whisked Margaret off to Winchester for the summer of 1886, from where she wrote Septimus a love letter asking forgiveness 'for the way I forced myself on you,' and confessing that 'I always feel so strongly attracted towards you, and have done for the past three years.' The letter seemed like a good idea at the time, but she instantly regretted it. 'How could I write such a letter – what would he think of me?' she asked herself.

And all before she set foot outside the British Isles.

There were further exchanges in which Septimus seemed decidedly underwhelmed by Margaret's attentions; a fact compounded by him being spotted walking arm in arm with another girl. He was also revealed to be a drunkard, and Margaret describes how a 'dreadful horror' seized her as she 'could not disguise from myself that the vice of intemperance was growing upon him daily.'[7] Septimus had also run up debts, and in the spirit of the Victorian cad which he was increasingly turning out to be, he did a moonlight flit – probably to escape his debts, but quite possibly also Margaret's suffocating love.

In 1889, when Margaret was twenty-seven, she inherited a share of a £30,000 bequest from her uncle, which allowed her to claim that hallowed financial status 'Of Independent Means.' It was also a bit of a coup for Septimus. Margaret had by now hunted him down to his native Limerick in Ireland, where she pursued him with the same enthusiasm as she had in Norwich. Unsurprisingly, Septimus now seemed a lot more interested in Margaret and her 'means,' writing to her that 'now you are independent, I should be very delighted to see you' – of course he would be! Within seven days they had met in Limerick, had several episodes of tender kissing and as far as Margaret was concerned, she was as good as engaged. Further heartache was to come, however.

The couple resolved to tell Margaret's mother of their plans, and it was at that point that the love story started to unravel. Margaret returned to England and announced her engagement to Mamma, an episode one suspects she rather plays down in her diaries, saying her mother was merely 'surprised' and that 'it will take a time before she gets accustomed to the idea.'[8] No doubt she was utterly horrified by the turn of events, but rather sensibly tells Margaret that Septimus must write to her with an explanation and an idea of his financial situation.

This alone was enough to put Septimus off for good, and although Margaret had kept her side of the agreement, Septimus was never seen or heard from again. His aunt Eleanor in Limerick, whom Margaret had met on her visit, took a similarly dim view of proceedings to Margaret's mother. Her damning testimony of her nephew, which she wrote in a letter to Margaret in 1890 read, 'I think the very best thing you can do is to banish all thought of him from your mind. He is not in any way worthy of you and I scarcely think him capable of caring much for anyone but himself. It grieves me to have to say this but I should like to open your eyes to his defects.'[9]

'It's as bad as it could be,'[10] Margaret recorded in her diary.

She never forgot Septimus. When her box of diaries was opened in 1978, it contained a letter – written in the last year or so of her life - addressed to whoever would open the box. It read:

> The greatest passion, and perhaps the most noble love of my life, was no doubt for Septimus Hewson, and the blow I received from his heartless conduct left a scar upon my heart, which no length of time ever quite effaced.[11]

Poignancy aside, the heartache Margaret experienced during her formative years was beneficial in one respect - Septimus's rejection provided the catalyst for her travels. To mend her broken heart, she needed distraction, and her growing interest in butterflies and her desire to escape the stuffy familiarity of provincial life led her to take some bold decisions – namely that she was destined for a life of entomological discovery and endeavour. She may not have been able to catch the love of her life, but she was determined to net a collection of butterflies and become an authority in her chosen field of study, lepidoptery.

Margaret's interest in butterflies was awakened on a trip to Switzerland, where she would spend her afternoons butterfly hunting, 'a pursuit which once started soon became all absorbing.' 'I was a born naturalist,' she wrote, 'though all these years for want of anything to excite it, it had lain dormant within me.'[12]

Between 1891 and 1900, Margaret visited Switzerland, Sicily – where she became the first British butterfly collector to brave the south Italian brigande – Austria, Germany and Hungary. Her preferred mode of transport was that great nineteenth-century female liberator, the bicycle, which she 'soon caught the spirit of,' and used throughout her adventures. Romance was never far from her mind, and during this period she managed to either fall in love with or reject an opera-teacher in Milan, an ageing professor on the Italian Lakes, a 'tall Italian with a dark beard, who had shown me some little civilities, such as lending me his Baedeker,'[13] a hotel proprietor's son, a French hotel guest who tried to visit her bedroom at night, a 'young and rather good-looking'[14] Italian youth, a Hungarian doctor, and finally, a Greek guide. Frankly, she must have been exhausted.

In 1899, she spent the butterfly season in the French Alps and was thrilled to finally meet her idol, Mr Henry Lang, whose reference book on European butterflies she used for her studies.[15] Margaret was by now an experienced lepidopterist and collected caterpillars as well as butterflies so she could breed her own perfect specimens; her diaries and notes show a serious approach to scientific study and she was able to successfully establish the best conditions to produce the healthiest specimens. Margaret's hard work paid off when Mr Lang visited her on her return to England. She wrote of his visit that Mr Lang 'almost made me conceited by his lavish praise of my collection, and all that I had done since I had become an entomologist. Just now I was 'booming' in the Entomological World.'[16]

Having explored what butterflies Europe had to offer, as well as what men, Margaret needed something a little more exotic to add to her fledgling butterfly collection. In 1901, her search led her to Damascus, where she was wooed by a Syrian dragoman called Khalil Neimy who became not only her travelling companion but one of the great loves of her life. He was twenty-four, she was thirty-nine. He was also married, but this seems to have been of little consequence to Margaret who was clearly of the 'you can't choose who you fall in love with' school of thought. 'I often reproached myself for Neimy's attachment to me,'[17] Margaret wrote in her diary, 'but how could I help it? The man swore he had no wish on earth except to make me his wife. I didn't care a damn about him. But I began to find his untiring devotion and constant adoration decidedly pleasant…More than once he would say: 'If you will not marry me I will follow you all over the world wherever you go, I will never leave you.''[18] It all sounded very familiar, except this time it was Margaret's turn to be suffocated to death by hopeless adulation. Margaret 'felt guilty to think what a boy he was' and reminded him of the age gap between them hoping it would dissuade him. 'But I was wrong,' she confessed. 'And then – out there beneath the shadows of those great rocks near Baalbek, on that glorious summer morning I solemnly vowed to Khalil Neimy that I would be his wife…I kissed him on the cheek, which was smooth and pink like a boy's, and then we held each other's hands and swore to be true. And all the time the big, brown butterflies flitted unmolested to and fro among the hot rocks.'[19]

Over the next twenty-eight years, until Khalil's death in 1929, the two worked and travelled together, first to Asia Minor (now modern-day Turkey) then to South Africa and Rhodesia, and across the Atlantic to the USA, Central America and the Caribbean. In between trips, Margaret wrote up an account of her travels and her scientific research, and wrote articles for *The Entomologist* and *Transactions of the Entomological Society*, which her academic peers praised for her observational astuteness.[20] Her work was always accompanied by exquisitely detailed illustrations, and she kept her own collection of sketches and watercolours, which were widely regarded as excellent examples of scientific artistry – so much so that they are now held in the Natural History Museum's collection and are regularly displayed in exhibitions. Her lectures both at home and abroad were well attended and in 1912 she was invited to join the prestigious Linnean Society, the world's oldest natural history society.

Before the Great War, Margaret was able to fit in a visit to India, Ceylon (now Sri Lanka), Nepal and Tibet which yielded more watercolour illustrations and fresh additions to her vast collection. Her last entomological expedition with Khalil was to the Philippines and following his death in 1929 she confided to her diary that her only source of comfort was her caterpillars. Alone, and now in her mid-sixties, Margaret continued exploring during the 1930s, visiting West and East Africa, Indo-China, Hong Kong, the Malay states, Brazil, the West Indies and finally Trinidad, where she died doing what she loved most. In the fifty years that Margaret had spent travelling she amassed a collection of 22,000 butterfly specimens from sixty countries, across six continents.

Her diaries may read as a curious mix of entomology and Mills and Boon-esque scenarios but to take her personal narrative at face value is to devalue the extent of her work and achievements. Her diaries only highlight one facet of an exceptionally talented and complex individual, who has become marginalized in the scientific community thanks, ironically, to her own conscientious diary keeping.

The tone was set shortly after the opening of the sealed box in 1978, when Margaret was promoted - by the very museum she had bequeathed her collection to - as a 'girl in love,' whose passions were 'crippled by Victorian morals' leading her to seek refuge in the pursuit of butterflies.[21] Her collection was also deemed 'not outstanding.' The narrative that followed was one of 'eccentric spinster goes butterfly hunting to escape the confines of her sex.' From the perspective of the reader, the joy of following her experience of love, and love-lost, is inescapable and it helped to sell the books that her diaries would eventually be edited into. Yet to focus purely on her romantic endeavours is to suppress her role as professional scientist at work. The men she met and her relationships with them were really only a pleasing by-product of her travels and expeditions. Butterflies were Margaret's first love and only real interest, and to lose sight of that is to do her a disservice. As her contemporary, lepidopterist Norman Riley wrote in 1940: 'Her great passion, however, was collecting butterflies, an interest which she first developed in about 1883, and which from then onwards led her every year further and further a-field in search of material for her collection.'[22] The men may have come and gone, but once caught and pinned, the butterflies weren't going anywhere. Her collection speaks for itself, and it is around that that the narrative of her life ought to be constructed, rather than her legacy of love.

Lady Jane Digby also knew something of love. Blessed with beauty, she found herself roaming abroad in order to escape a scandal of her own making at home.

Born in 1807 at the family seat of Holkham Hall in Norfolk, Jane was highly educated and, as it would transpire, highly sexed – a dangerous combination in nineteenth-century high society. Her grandfather, Thomas William Coke, encouraged Jane to be politically and culturally precocious, as well as a skilled equestrian - assets that would allow her to flourish in an enlightened post-Regency world.

With such a glittering résumé, Jane was an extremely marriageable prospect and in 1824, at the age of just seventeen, she became the wife of Edward Law, 2nd Baron Ellenborough (later Earl of Ellenborough). It was not a love match and Jane sought excitement and romance elsewhere, notably in the bed of her cousin, Colonel George Anson. In February 1828, Lady Jane gave birth to a son, Arthur Dudley Law, whose paternity was later brought into question when details of the full extent of Jane's extra-marital affairs emerged. As was customary at the time (and still is) Lady Jane's dirty laundry was fully aired via the national press; made all the juicier by the involvement of a Bohemian nobleman and visiting diplomat, Prince Felix of Schwarzenberg. The story rumbled away during the late 1820s before finally reaching its natural conclusion in 1830, when the cuckolded Baron Ellenborough was granted a divorce on the grounds of his wife's flagrant adultery – which had recently been made very obvious when she gave birth to Prince Felix's illegitimate daughter Mathilde in 1829. Further evidence of Jane's adultery, which unfortunately for her was abundant, was provided by a succession of witnesses, including servants who testified that bed sheets had been 'rumpled' when they shouldn't have been, and that the Prince had been seen 'lacing her stays.'[23] It was *the* aristocratic scandal of the day and made for very distasteful reading, although no doubt it was devoured with relish by the rest of society.

To her credit, Jane was nothing but candid with her own revelations: 'Lady Ellenborough then confessed to her [the governess] that she was with child, and that the father of the child was Prince Schwarzenberg, who had drunk tea, and passed the night with her, and part of the next day at Brighton.'[24] She could do no less given that she was expecting a second child with Felix at the time of the divorce proceedings.

In the wake of her divorce absolute, Jane decided to leave England permanently for the continent. Over the next twenty years she travelled slowly eastwards, leading a life that was notorious even by the louche Regency standards of the day. Her conquests included Ludwig I of Bavaria and Baron Karl von Venningen, whom she married in 1833. Two more children followed before Jane found a new love, Count Spyridon Theotokis of Greece. This affair was the catalyst for more scandal when her spurned husband Baron von Venningen challenged the usurper Theotokis to a duel, wounding him in the process. In an act of remarkable charity given the circumstances (or guilt, perhaps), Venningen released Jane from their marriage *and* took care of their children, and the two remained friends throughout their lives.

After a quick conversion to the Greek Orthodox Church, Jane married Theotokis in Marseille in 1841, the year after she had given him a son – Leonidas, who tragically died after falling from a balcony in 1846. The couple moved to Greece but divorced that same year. Jane then became the lover of Greece's King Otto, the son of Ludwig I of Bavaria, Jane's former lover from a decade earlier, before moving on to one of Otto's generals, national hero Christodoulos Chatzipetros.

By contemporary standards, Jane's most shocking dalliance was yet to come. It was 1853 and Jane, now aged forty-six, had made her way as far east as Syria, à la Lady Hester Stanhope. Here she would meet and fall in love with Sheik Medjuel el Mezrab, whom she had enlisted to escort her caravan through the desert. Jane had been away from the stuffy Regency drawing-rooms for long enough to not care a fig for propriety, class or convention. The couple married under Muslim law and Jane took the name Jane Elizabeth Digby el Mezrab. During their successful twenty-eight-year relationship they established an unconventional lifestyle, which, most importantly, accommodated each other's needs: Jane's to have access to some semblance of 'society' albeit an alternative one, and an occasional roof over her head, and Medjuel's to maintain his nomadic lifestyle within the brotherhood of his desert tribe. Their solution was simple. Jane built a palatial villa in Damascus, where they spent half the year, and the other half was spent under a goat's hair tent with the rest of the Bedouin tribe. Jane learned Arabic and adopted Arab dress, and Medjuel agreed to monogamy, despite the customs of his faith. Jane never converted to Islam. Instead, their relationship stood as a cultural bridge between the Christian and Muslim worlds.

Her friend and confidante, Isabel Burton – herself a renowned explorer-writer and wife of explorer-extraordinaire Sir Richard Burton – took it upon herself in 1873 to lay some of the later scandals surrounding Jane to rest. Isabel's defence of her friend was made under the misapprehension that Jane was dead, whereas, in fact, she would live another eight years after Isabel's attack on the *German Gazette* for their premature obituary, which Isabel felt was littered with falsehoods. The grieving Isabel wrote:

> Knowing that after her death all sorts of un-truths would appear in the papers very painful to her family – as, indeed, she was not spared while living – she wished me to write her biography, and gave me an hour a day until it was accomplished. She did not spare herself, dictating the bad with the same frankness as the good. I was pledged not to publish this until after her death and that of certain near relatives…It must have come from a very common source when such English as this is used: 'between Beyrout [sic] and Damascus she got pleased with the camel driver!'[25]

Isabel went on to describe her friend, and her relationship with Medjuel in the most glowing terms. She wrote: 'On the journey the young Shaykh [Medjuel] fell in love with this beautiful woman, who possessed all the qualities that could fire the Arab imagination…It ended by his proposing to divorce his Moslem wives and to marry her; to pass half the year in Damascus (which to him was like what London or Paris would be to us) for her pleasure, and half in the Desert to lead his natural life.'[26] The arrangement suited Jane for it allowed her to make real the 'romantic picture of becoming a queen of the Desert and of the wild Bedouin tribes' - which fitted Jane's wayward Byronic profile perfectly. And when she'd had enough of sleeping on the ground, she could escape to her Damascus bolt hole and become the society hostess she was always destined to become – on her own terms, of course.

By August 1873, the English press was redacting the false report of Jane's death. The rumour had, allegedly, been put into circulation by one of Jane's 'bitterest enemies in the neighbourhood of Damascus, which she had made for some years her headquarters, opening her Eastern home to all sorts of visitors from the West. As well as to her Oriental friends.'[27] Drama never really stopped following Jane around; she was good fodder for filling the column inches; and despite settling several thousand miles away from her native homeland, her 'singular career of adventure'[28] was still considered newsworthy some twenty years after it began.

Even her funeral, when it did finally happen in 1881, had a melodramatic air to it. Jane Digby el Mezrab died of fever and dysentery on 11 August and as a 'good Christian lady' – as described by Isabel Burton in her premature obituary – she was to be buried in Damascus' Protestant Cemetery. Christian funereal practice dictated that the bereaved Medjuel travel in the first carriage of the funeral cortège, but the poor Sheik was so overcome with grief that he hurled himself out of the carriage and ran in the opposite direction to the melancholy procession. As is customary amongst the English when faced with a potentially embarrassing situation, the other mourners decided to ignore this eccentric display and continued with the service as if nothing had happened. As the clergyman began to sprinkle earth on the coffin, the Sheik galloped into the cemetery on Jane's favourite black mare. After a moment of silent reflection by her graveside, he galloped back out to the desert – no doubt to find the solitude he needed to commemorate her in his own way, amongst the comfort of his own people.

Jane's was a life well lived, even if it was a little unconventional for the time. Such was the fascination for what the *Birmingham Mail* referred to as her 'madly wayward career,'[29] that comparisons were inevitably drawn with contemporary works of romantic fiction. Swiss-born writer, Victor Cherbulliez's *Miss Rovel* (1875) was said to resemble something of Jane's life; the 'curious French novel,' described Jane Digby, 'as only a Frenchman dares to describe a lady's character,' which we can only assume means it was a risqué account. *The Cheltenham Examiner* chose to reference French novelist Edmond About in her obituary in September 1881. The writer had allegedly used Jane's story as 'a subject of romance. The precise character we do not profess to know, but our impression is that it must be more sensational than edifying.'

Lady Jane Ellenborough, as was, also made a fictional guest appearance in the *Two Admirals* – a little known work by the even lesser-known writer, Mr Wyndham. The novel, which the *Alcester Chronicle* reported was, 'still sold at the railway stations,' (sounds promising) contained a 'not exaggerated description of Lady Ellenborough, and of her riding her ponies barebacked in the grounds of Roehampton. It is one of Mr Wyndham's best and most pathetic tales.'[30] Jane also appeared as herself in Lady Anne Blunt's first account of her own travels to the Middle East to buy Arabian horses; and in her friend Lady Isabel Burton's *Inner Life of Syria, Palestine and the Holy Land* (1875). That she was immortalized in this way would no doubt have pleased Jane who garnered a sort of celebrity status, even before the term came into popular usage. The 'half-forgotten scandal' – to steal the headline of *The Cheltenham Examiner* - of her life is still entertaining the masses today. At the time of writing, her character appears in a new BBC adaptation of Jules Verne's *Around the World in 80 Days*, as the ultimate aristocratic adventurer par excellence.

Her remarkable life seemed to come straight from the pages of a book and the *Berwickshire News* were spot on when they wrote in October 1881: 'The death is announced at Damascus of lady whose extraordinary career is another illustration of the Byronic adage that truth is stranger than fiction.' In Jane's case, it most certainly was.

Lady Hester Stanhope, who had trodden the dusty desert path to Syria a full four decades before Jane, had similar rebel tendencies and an equal capacity for romantic adventures and outlandish escapades. Born in 1776 at Chevening in Kent, the Palladian pile of the Earls Stanhope, Hester was motherless by the age of four and spent her childhood being passed around various relatives. In 1803, she went to live with her uncle, the Prime Minister William Pitt, and carved out a role for herself as his glittering hostess. With her witty repartee, intelligence and political credentials, she was an asset to any dinner party or aristocratic soirée. Toasts were proposed to her beauty, and much was made of her 'magnificent and majestic figure'[31] (she was almost 6ft, which was extraordinarily tall for a woman of the time) and the way 'roses and lilies were blended in [her] face.' She 'diffused happiness' all around her.

Hester came alive in the company of men, and they in turn, worshipped her. Future army general, William Napier, remarked at the time that 'it was quite impossible not to fall at once into her direction and become her slave.'[32] At her uncle's residence Walmer Castle, she found herself 'exactly in the sort of society I most like.' 'There are generally three or four men staying in the house,' she informed a friend, 'and we dine eight or ten almost every other day. Military and naval characters are constantly welcome here; women are not, I suppose, because they do not form any part of our society. You may guess then what a pretty fuss they make over me.'[33]

When Uncle Pitt died in 1806, Hester was deprived of the influence and power she had hitherto enjoyed. She was still wealthy – Pitt had arranged for Parliament to grant her an annual pension of £1200 – but she lacked purpose. She developed a close personal relationship with a senior soldier in the British Army, General Sir John Moore, but he was killed in the Battle of Corunna in Spain in 1809 – tragically, alongside Hester's 'dearest, delightful, amusing' half-brother, Charles, who was shot in the heart by a bullet whilst he was congratulating his men in the 50th Regiment for their efforts. It was a double emotional blow and more than likely the catalyst for the eccentric overseas wanderings that she would later become known for.

Of course, Hester's emotional state of mind wasn't the *only* catalyst for her travels. Her very nature meant she was predisposed to a life of adventure. During her childhood years Hester was frequently cast as the 'wilful' and 'rebellious' one - her grandmother, Grizel, confided to a friend that 'Hester is quite wild. I am forced to send assistance from here to keep her within bounds.'[34] She also showed an early interest in travel. When she was on a family trip to Hastings beach at the age of eight, she managed to slip away unnoticed, commandeer a boat and row herself out to sea. Her destination was France. Hester later claimed that she had not been frightened by the experience, just amused by the look of horror on her governess's face. This devil-may-care attitude was Hester's approach to life until the day she died – she was completely unperturbed by rules, conventions or behaving 'appropriately,' so it is unsurprising that in one of the darkest periods of her life, she sought escapism through travel and exploration.

Hester left England in February 1810, and never returned. Among her entourage of adoring followers was physician Dr Charles Meryon, who would later publish her memoirs, and two maids, Elizabeth Williams and Anne Fry, whom Hester had known since her days at Chevening. She went first to Gibraltar, where she took a young lover – professional traveller Michael Bruce, who at twenty-three was eleven years her junior, and whose spirit of discovery and adventure was more than a match for Hester's.

After a quick trip to Malta, Hester decided to travel further east. Arrangements were made, which included scaling down their considerable luggage: 'My baggage will amount to a dozen shirts, my cot and my saddle; Lady Hester's and Mr Bruce's on the same scale. All useless luggage is to be left behind.'[35] With only their 'useful' luggage in tow, the group left Malta and headed first for the island of Zante, then on to Patras on the Peloponnese peninsula, finally arriving in Athens on 12 September 1810. Here they met Romantic movement royalty, the 'mad, bad, and dangerous to know' poet Lord Byron. This meeting of the two disruptors did not go well. At a dinner, Hester and Byron debated fiercely on matters of equality. Byron's account of their meeting was particularly scathing. 'I saw the Lady Hester Stanhope at Athens,' he wrote in a letter, 'and do not admire ''that dangerous thing a female wit''...I have seen too little of the Lady to form any decisive opinion, but I have discovered nothing different from other she-things, except a great disregard of received notions in her conversation as well as conduct.' Hester's opinion of Byron isn't recorded but there was certainly no love lost between the two, which suited Hester very well since she still only had eyes for Michael Bruce.

Hester travelled onwards to Constantinople, renting a luxurious apartment and making full use of the local Turkish baths. Here in a steam filled chamber, she would lie face down on a marble slab and be scrubbed, pummelled, massaged, and wrapped in cloths. After a quick doze, she would wake up refreshed and hungry.

Her visits to the *hamam* provided Hester with a much-needed break from her younger lover, who was becoming increasingly irritating. Bruce had asked Hester to marry him. She had politely declined. So he took himself off on an expedition to Persia without her – no doubt sulking all the way. In the meantime, Hester self-medicated her frustrations by smoking *chibuk* and *nargileh* pipes, filled with a mixture of tobacco, opium and rose-scented water; and visiting the harem, where even she was shocked by the lewd cavorting; commenting that the dancing was the most disgusting thing she ever saw.

With the return of the lovelorn Bruce from his travels, the party decided that Egypt and Syria were to be their next destination. They left Constantinople by boat on 23 October 1811, making it as far as Rhodes where they picked up fresh supplies before they set off again for Alexandria.

They made it halfway there before a storm blew in and the party was forced to turn back to Rhodes. The Greek *caïque* (a small, wooden fishing boat or skiff) they were travelling in sustained damage and the boat started to take on water. To everyone's relief they made it to within a mile of the Rhodes coastline but by then the boat was so flooded that the order was given to let down the long boat and abandon ship. With the storm still raging overhead, all twenty-seven passengers managed to cram into the long boat, but the chances of survival were slim and rapidly diminishing with every roll and swell of the boat. Hester's beloved dog refused to jump from the caique, and she sobbed as she watched the sea claim its first victim.

A retrospective of her adventures thus far appeared in the press in 1816. 'Lady Hester Stanhope,' it was reported, 'merits a place among the most celebrated and intrepid travellers of the present age.' The article, which was syndicated across Britain and its colonies, claimed that 'cast on a barren rock, she seemed to be destined to perish of hunger; but an English ship, which appeared on the following day, took her on board, and conveyed her to Syria.'[36] Following her death in 1838, the same incident was described in the following way; '...the ship was wrecked, Lady Hester Stanhope's property was all lost, and it was as much as she could do to save her own life. Nothing, however, could shake her resolution. She returned to England, gathered the remainder of her fortune, [and] sailed again for Syria...'[37]

In fact, neither of these reports were entirely true. The crew managed to steer the longboat onto a rocky outcrop, where they flung themselves onto the rocks, bruised and battered from their ordeal. The captain then decided that some of the crew should row ashore to get help, leaving everyone else on the rocks to weather the storm, which had now abated slightly.

They finally returned thirty hours later with provisions, and after everyone had had their fill, as well as a generous amount of alcohol to fortify themselves, a second landing on the coast was attempted. After doing battle with the Levantine Sea for four hours, a violent wave finally threw the hapless party onto a stony beach. They then faced a day's trek to reach the nearest village.

Hester had managed to cling on to a small tin, in which she had secreted General Sir John Moore's bloodstained glove and some other personal effects, and Bruce had a small amount of money tied around his waist, but everything else, including silks, weapons and jewels, had perished – claimed by the ocean as a hard won prize.

The incident, whilst terrifying, marked a defining moment in Hester's adventures abroad. Rather than dwell on all that she had lost, she chose to see the shipwreck as the point at which her relationship with her past was severed for good. It was a new beginning. And a new beginning meant a new look.

Ottoman custom dictated that as a woman, Hester should be covered and veiled. But she shunned this idea on the basis that if she identified as a Turkish woman, she wouldn't be allowed to be seen conversing with men. The sensible compromise was to dress in the style of a Turkish man instead, so she decided to wear silk shirts, a waistcoat and breeches and topped it off with a colourful turban. The *Leeds Intelligencer* remarked that a traveller to the region 'would behold her dressed in Oriental garments:'

> Her head was covered with a turban made of a red and white cashmere. She wore a long tunic, with open loose sleeves; large Turkish trousers, the folds of which hung over yellow Morocco boots embroidered with silk. Her shoulders were covered with a sort of burnous [a hooded Arab cloak], and a yataghan [a long knife] hung to her waist. Lady Hester Stanhope had a serious and imposing countenance; her noble and mild features had a majestic expression, which her high stature and the dignity of her movements enhanced.[38]

The 'new and improved' Lady Hester Stanhope cut a magnificent figure. She had also shaved her head, although this was more likely to be a measure to lessen the risk of lice infestation than for the sake of trendsetting or an act of rebellion. Her impressive defiance of all conventions won admiration all round, particularly with the local Turks. With her new found confidence, Hester set her sights on travelling to Palmyra and visiting the ruins of Queen Zenobia's city in the Syrian desert – a journey fraught with danger because it involved crossing Bedouin territory.

Over the next two years, Hester travelled the Ottoman world, forging friendships and new alliances that would allow her safe passage over the tribal badlands. She also gathered people about her – whether by accident or design – and her entourage grew exponentially. Hester and Bruce each had a personal servant, valet and interpreter who went with them everywhere, a cook, two grooms for their five horses and two men to pitch tents.

With an impressive retinue of admirers, adorers and servants, Hester set about conquering various parts of Egypt, Palestine and Syria. When news reached the local people in Jerusalem that an 'English princess' was in residence, a mob camped outside her quarters hoping to be showered with gold. But this was nothing compared to the welcome she received in Palmyra in 1813. Hester had longed to see the once-magnificent city, which was now reduced to ruins, seeing herself as a modern-day Zenobia, leading her faithful followers to triumph. As the first Western woman to see the ancient city, the welcome she received from the Sheikh of Palmyra did not disappoint. Three hundred of his warriors were orchestrated to 'attack' Hester and her caravan – a customary part of tribal etiquette, to which Hester was expected to mount a similar response. Once everyone had proved what brilliant and brave warriors they were, Hester processed through the legendary colonnaded streets, followed by singing flower-girls and musicians playing Arabian instruments and singing odes. Brimming with self-importance Hester declared later that, 'I have been crowned Queen of the Desert under the Triumphal Arch at Palmyra…and, if I please, I can now go to Mecca alone – I have nothing to fear. I shall soon have as many names as Apollo. I am the sun, the stars, the pearl and the lion, the light from heaven, and the Queen.'[39] Modesty was not one of Hester's strengths, and it is unsurprising that an early twentieth-century analysis of her exploits asserted that 'she wanted to reign, to rule, to govern, and as that was no longer a possibility in the political world, she resolved upon seeking some new sphere where she would always be first.'[40]

For all her bragging, Hester really had achieved something extraordinary. She was a woman – not just a woman, but an aristocratic Western woman – penetrating the very heart of a war-torn Middle East, whilst navigating a dangerously shifting tribal and political landscape, at a time when other ladies were content to sew by the hearthside and sit on the Board of the local poor house. Hester's idea of *noblesse oblige* was to give protection to hundreds of terrified refugees during the siege of Acre in 1832 – all of whom needed to escape the fall of the Ottoman controlled countries of

Syria and Palestine to the Egyptian army, and the inevitable round of beheadings that were destined to happen as a result.

By then, Hester had her own 'fortress' – the remote abandoned monastery of Dar Djoun in the foothills of Mount Lebanon, near Sidon. Bruce was long gone, to find new adventures and women in other parts of the world and Hester had no one to please but herself, so she wasted no time in fitting out her fortress to her own specifications. A description of Hester's hilltop residence was published by the *Belfast Commercial Chronicle* six months before her death in 1839. The article, titled 'A Visit To Lady Hester Stanhope,' was written 'By An American' who remains anonymous but was certainly someone Hester already knew or was interested in meeting – she chose her guests and visitors wisely and only a select few were permitted a reception with her, particularly in her later years. 'We came in sight of the little insulated mountain, where Lady Hester Stanhope had established her lonely dwelling,'[41] the American wrote:

> It is almost conical, and separated by a deep valley from the other hills. We toiled up its precipitous side by a narrow winding path, enjoying the full benefit of a Syrian mid-day sun. When on the top we stopped a moment to rest, and to survey the prospect around us. Steep valleys on every side seemed to enclose similar hills...The house, or rather the cluster of houses, is built in the Arab manner, low, irregular, and almost detached. It is of stone, rather rudely constructed, and surrounded, as is usual here, with stone wall. There are some fig and pomegranate trees, vines and flowering shrubs, cultivated with care, and furnished with water brought from some distant spring in the valley below, upon asses for the hill itself as destitute of water as the deserts of Arabia.[42]

Hester, who was now sixty-three, received her guest 'with great kindness,' still dressed 'like an Arab, clothed in white, with a turban upon her head, and smoking a long pipe.'[43] The writer described her as:

> ...tall and spare, with a worn and sickly complexion... There was a settled melancholy, which added to the interest of her appearance; and the recollection of what she had been, contrasted with what she was, produced a powerful impression on each of the party.[44]

Written just months before her death, Hester's charismatic light was fading fast. She had run up vast debts to maintain her exotic A-list lifestyle, leading the British government to divert the annual payments bequeathed to her by Pitt straight to her creditors. However, despite her pecuniary difficulties, Hester was still aware of her social responsibilities. She became a generous de facto ruler of her own little hilltop kingdom and the local people testified to her kindness to those most in need. It was to the refuge of her stone eyrie that the local people fled in 1832 to escape Ibrahim Pasha's army. And Hester didn't disappoint – she offered shelter to men, women and children, 'stripped to the bone' and 'idiotic from fright.' Some were missing eyes and body parts; others had been completely lost in the melee. Hester did what she could, including borrowing more money so she could provide for everyone, but many died and were buried in a mass grave. She was comforted by the knowledge that she had saved as many people as she humanly could:

> ...my house was like the Tower of Babel, filled as well as the village, with unhappy people from Acre of all nations, but with the blessing of God I got through it all, and saved many doomed to have their heads cut off.[45]

The people had looked to Hester, their 'Queen of the Desert,' for protection – and they found it. But she was more than just a matriarchal figure who tended the injured and destitute, she was also capable of inspiring the same level of deference and respect that the Arab chieftains and overlords would extend to their equals. She operated in a man's world but found herself more than a match for the opposite sex. Brave to the point of recklessness, the wilful and at times baffling Lady Hester Stanhope would, tragically, have a rather pathetic end to her story.

She became something of a hermit in her final years, existing in a haze of incense and rattling around her residence in a cloud of *chibouk* smoke. She would eventually die alone in her bed in June 1839, most of her servants having left her – either because she hadn't paid their wages or due to her increasingly erratic behaviour. Those that stayed with Hester until she took her last breath pinched some of her meagre possessions before they left. She received a Christian burial; her final resting place was the vault in her garden which already contained the bones of a French general. That was her wish. Why Hester wanted them to share a grave was a mystery that went with her into the vault.

Her loss was lamented in Britain – not out of any real sense of grief but because she had kept the nation amused with her notorious exploits over the years. Of her downfall and subsequent death, the *Leeds Intelligencer* wrote:

> …she was left the title of Queen, but it was but an empty name; a mere recollection, and again the monastery's silence ruled over the solitude of Djouni. A Queen, stripped of her glory of a day, Lady Hester Stanhope has expired, the sport of fete, at the moment the East is convulsed. She has expired in obscurity and loneliness, without even mingling her name with the great events of which it is now the theatre.[46]

It was for love - lost and found - that Hester travelled, but really her one true love was her adoptive country, Syria. That, and a love of herself – a love that meant she could recognize her own need to flout convention and be who and what she wanted to be, regardless of what anyone else thought. And that's as good a legacy as any.

If Hester Stanhope was committed to discovering herself through her travels, Lady Jane Franklin was desperate to discover the whereabouts of her husband. Sir John Franklin had disappeared on an expedition to traverse the Northwest passage sometime after 1845, and Lady Jane would not rest until she knew his fate. It took her twelve exhausting years and seven sponsored expeditions to find out the truth, but by sheer strength of will – and limitless funds - she did eventually succeed.

Jane's adventuring career started young. As a child she accompanied her father on several trips abroad, diligently recording her travels in a journal and taking inspiration from her fictional heroine, Emily St. Aubert – the energetic and resourceful Gothic protagonist in Ann Radcliffe's *The Mysteries of Udolpho*, which Jane read and re-read several times during her youth. Like many young aristocratic ladies and gentlemen of the day, she finished her education with an immersion in the cultural delights of a Grand Tour.

At the age of eighteen she was already proving to be a keen climber and notched up several mountain peaks, including the 3000ft Cader Iris, legendary Welsh seat of King Arthur, and Mount Snowdon - an impressive 3,560ft, which made it the highest peak in Wales. She added Montanvert, one of the three large glaciers on Mont Blanc and the highest peak in Europe (15,781ft) to this list in 1814. When she wasn't trekking up mountains, she also enjoyed intellectual pursuits, allegedly reading nearly three-hundred educational books over the space of three years, and frequently attending lectures at the Royal Institute.

On 5 November 1828, Jane married Royal Navy officer and Arctic explorer John Franklin. She was a month from her thirty-seventh birthday and the widowed Captain Franklin was forty-two with a five-year old daughter from his first marriage to the poet Eleanor Porden, who had been one of Jane's closest friends.

Now the wife of a serving Navy officer, Jane had the perfect excuse to visit more exotic and far-flung locations than she had previously had the opportunity to travel to. Wives were not permitted to live on board, but Franklin's posting to Malta in 1829 meant that Jane could explore the Mediterranean and intermittently visit her husband. From Malta, Jane sailed to Corfu and then took a steamer along the coast of North

Africa, marvelling at a set of new sights in what was her first experience of a non-Christian world. There, under the hot sun, she wondered at the white walled buildings of the casbah, and the minarets perched on the mosques; as well as attempting to avoid several undesirable consequences of her location, such as mosquitoes, plagues of strange insects and unfamiliar and not always pleasant smells. By 1832 she was embarking on a trip to Alexandria and the Holy Land, treading in the footsteps of the mighty Hester Stanhope, and trying to evade the worsening political situation. However, unlike Hester, Jane was only really interested in riding donkeys, scrabbling over ruins and filling vials of water from the sacred Jordan, rather than becoming embroiled in political machinations and seeking self-aggrandizement.

Jane was cut from a different cloth to Hester – both were intent on satiating their need to explore, as well as pursuing love, in Jane's case her love for her husband, but there the similarities ended. Where Hester desired status and recognition amongst her peers, as well as her trump card – the ability to shock and surprise everyone around her - Jane's endeavours were a little more low-key. Yet, it would be wrong to consider them less exciting. Jane's pleasure derived from experiencing the ancient atmosphere of a place. She had an innate appreciation of the cultural, religious and historical significance of the places she visited; and to breathe in the very air they inhabited was all the stimulus she needed. But soon, even that wasn't enough. Jane was certainly aware of Hester, and no doubt admired her travelling achievements, but Hester had crossed the invisible line that divided propriety and indecency - a fate that Jane did not wish for herself. As the wife of a naval officer, and a 'lady,' Jane found that doors opened more freely for her than they would have done had she remained a spinster. Travelling 'alone' (servants didn't count) and unaccompanied by her husband was still a risky business – it was certainly unusual for the time – should she pack up and go home? Or push further on without her husband?

By 1833, Jane had finished exploring the Mediterranean, yet she wasn't quite ready to go home to England. Unlike Hester, she also didn't want to permanently settle where she had concluded her journey. Instead, she wanted to go even further around the globe. The other side of the world, in fact, to England's convict dumping ground – the Antipodes.

Fortunately, in 1836, the Franklin's received the news they had been waiting for – Sir John was to be posted to Van Diemen's Land (modern-day Tasmania) as the lieutenant-governor. On the evening of 26 August 1836, the Franklins, along with Jane's stepdaughter Eleanor and several other family members and close friends, set sail from Portsmouth – an occasion that warranted a special mention in the local paper: 'The *Fairlie,* for Van Diemen's Land, full of Emigrant's, will sail the first fair wind. Sir John Franklin, appointed Governor at that settlement, with his family, and Capt. Machonochie, R. N. who will act as his Secretary, and a large family, are embarked.'[47]

Travelling to what really was the 'back of beyond' in the nineteenth-century was not for the faint hearted. There was only one way to travel to the Antipodes, and that was by boat – a journey that took at least four months on average.

There was a reason Britain had been sending her felons there since 1787, and it wasn't for its proximity to other colonized countries. The original settlers from the First Fleet of two Royal Navy vessels, three store ships and six convict ships – over 1400 people in total – were sorely disappointed when they finally arrived at Botany Bay in 1788. The ships could not anchor close to the shore because the bay was too shallow, fresh water was scarce and the soil quality was poor. Since the point of the convict colony was to establish a new settlement and farm the land, a different location was hastily sourced – Port Jackson, now known as Sydney Harbour, which the first Governor of New South Wales, Arthur Phillip famously described as 'being without exception the finest Harbour in the World... Here a Thousand Sail of the Line may ride in the most perfect Security.'[48]

When Jane sailed into Hobart on the southeast corner of Van Diemen's Land in the first few days of January 1837, the settlement was already over thirty years old. The voyage had not been short of incidents. The ship had only made it thirty miles off the coast of southwest England when it had been caught up in a gale near the Scilly Isles. For three days and three nights, the ship had rolled with such intent that the passengers, which included three pregnant women and fifty-two children, had been unable to walk the decks without clinging onto the railings for dear life. One hapless passenger went overboard and had to be rescued, and two weeks into the journey a woman died leaving three children – one just seven months old. Two more passengers would die before the boat finally docked at its destination.

During a three week sojourn that the *Fairlie* took at the Cape of Good Hope, Jane found time to climb a mountain – this time it was Table Mountain - as well as learning how to handle a harpoon and dissect 'an enormous and disgusting looking fish called a squid.'[49] The Franklins arrival in Hobart was reported by the British press six months later, courtesy of a report filed by the *Hobart Town Courier,* which reported that, 'Sir John was loudly and repeatedly cheered as he passed along, and evidently appreciated the cordiality of his reception.'[50]

The island Jane now called home had a population of approximately 42,000 people, made up of 17,600 convicts, 24,000 settlers and 250 Aboriginal people, the majority of which resided in the two principal towns, Launceston and Hobart. A year after Jane arrived, the *Fife Herald* published 'Notes from Australia' – a suitably flowery account of an immigrant's first impressions of Van Diemen's Land on the approach to Hobart:

> The most striking of all is the bend of the river which forms the site of Hobart Town…The appearance of the rising capital from the deck is very picturesque, scattered as it is over a large tract of ground, irregular, yet pretty; and the churches, windmills, handsome houses, and elegant villas, which studded the river side, gave an idea of the importance and wealth of Hobart Town much beyond my most sanguine imaginings…[51]

Barring the local wildlife, which consisted of kangaroos, wombats, poisonous snakes and many undesirable species of insect-life, once a traveller strayed beyond the boundaries of Hobart, the island was a vast uninhabited wilderness.

To a woman like Jane Franklin, it would have been a veritable feast for her explorer-eyes. However, there was one minor problem: Jane was the wife of the lieutenant governor, and her official role included nothing more challenging than hosting tea and dinner parties and entertaining the wives of visiting dignitaries. If she had an opinion on anything, she was expected to keep her thoughts to herself.

Passivity had never suited Jane, however, and even before the *Fairlie* had docked in Hobart, Jane had done her homework, studiously examining reports and books regarding transportation and conducting lengthy discussions with prison reformer Alexander Maconochie – a fellow passenger on the *Fairlie*. What she really wanted was to be *useful*. The 'infant nation'[52] of Van Diemen's Land was now over thirty-years old and, in Jane's opinion, ripe for some intellectual cultivation. For Jane, it wasn't just a mission of educational improvement, it was a humanitarian effort – she embarked on her philanthropic projects in the hope that enlightening the settlers would broaden their own horizons and opportunities.

In the first two years of their residence, Jane also became the first European woman to travel overland between Port Phillip and Sydney, visiting the new settlement at Melbourne, and set her sights on becoming the first European woman to conquer Mount Wellington. Described by the *Fife Herald* as the 'the loftiest of the range…its clear, blue summit, its steep, craggy, and in many places perpendicular sides; its upper clefts lined with snow, and its base girdled with forests of Bush,' as far as Jane was concerned, Mount Wellington was there to be climbed, and she vowed to accomplish it by the end of 1837 – so she did just that. And she did it in style too, leading the party to the summit where everyone enjoyed an elaborate banquet. Not long after her ascent, a story appeared in the press recounting the mountaineering efforts of 'the 21st Regt. and their ladies,' who it seemed had tried to emulate

Jane's intrepidness by climbing Mount Wellington themselves. Unfortunately, they didn't take a guide, which meant they got lost for two days and were eventually discovered by the military in three separate groups, having argued over the correct route to descend the mountain. All three groups were eventually rounded up and described as 'a truly miserable spectacle…In all, this exploring expedition, we fancy, has quite satisfied each and all of the party.'[53]

In a final twist it transpired that the Surveyor-General of Van Diemen's Land, George Frankland, had been amongst the party. 'The most extraordinary part of it is this,' wrote the reporter, '[that] with all his knowledge of land surveying, trigonometry, and right lines, he could not discover what was a straight line; but he certainly did discover that it took from Friday morning to Sunday evening to form a circle, the extremes of which were Secheron and Mount Wellington.[54]'

In 1840 Jane founded, along with her husband, a state college known as Christ's College; the foundation stone was laid on 7 November of that year. She also established a 'Tasmanian Ladies' Society for the Reformation of Female Prisoners' in 1841 – her aim being to ameliorate the plight of the female convicts and broaden their horizons with the same activities she had enjoyed in her youth.

Next came a misguided scheme to rid the island of snakes. This involved Jane paying one shilling every time someone brought her a dead snake's head. This unusual enterprise was said to have cost her £600 (approximately £37,000 today) in one season and it had a disastrous impact on the productivity of the convicts, who would much rather earn an easy shilling killing a snake than working the land for a pittance – or nothing, in some cases. She eventually closed the scheme, but not before she had disposed of 12,000 snakes and a sizeable chunk of her own money.

The problem was, Jane was an outsider. She meant well but, like so many of her kind, she just didn't fit in. Jane was playing a risky game, meddling in matters that were conceived or resolved over a sherry and a cigar, and much slapping of backs. She was ill-suited to a life of gentility, sitting on the sidelines or in the shadows. The concept was anathema to her. She had intellectual resources and the wherewithal to deploy them – which in nineteenth century colonial society was a dangerous combination. Jane wanted to leave her mark, but the bar was set so low for women in terms of *doing something* – anything, really - that it rendered them invisible.

Yet unlike most colonial wives in Van Diemen's Land at that time, Jane's insatiable need to explore provided her with an outlet into which she could escape. She crossed six-hundred miles of rough and rutted dirt road to make the journey from Melbourne to Sydney – sometimes in a cart, but often by either riding a pony or walking – crossing water dykes and difficult terrain with stoic equanimity.

Despite John Franklin's requests for Jane to return to Hobart from her sojourn in Sydney, she did no such thing. She was having too much fun. Adventure would always trump all other things as far as Jane was concerned, although she was careful to package her exploits up as a 'research trip.' Her journals and letters to her husband included data on crops, the weather, manufacturing methods, and drainage, so that it would appear she was recording information that he would find useful. Thus, when John Franklin appealed for her to return to Hobart, she had no qualms in stating that she had no intention of doing anything of the sort.

But political in-fighting and the constant vilification of their actions in the press, both in Van Diemen's Land and back home in England, had begun to weigh heavily on the lieutenant-governor and his wife. Word of these unseemly shenanigans got back to England, and landed on the desk of Lord Stanley, the Colonial Secretary in Sir Robert Peel's Conservative government, who was unimpressed. Sir John Franklin was recalled in 1843 and the couple's Antipodean adventure abruptly ended.

The Franklins left Hobart on 3 November 1843 on the *Flying Fish* and arrived back on the sunny shores of England – Portsmouth to be exact – in early June 1844. There was a marked lack of interest in the return of the chastened couple. There would be no fanfare, cheering or military salutes this time. Instead, they got two lines in the *London Evening Standard*: 'Sir John Franklin, who, we understand, has re-signed the Governorship of Van Diemen's Land, arrived in town yesterday. We believe the successor to Sir John has not yet been appointed.'[55]

Less than a year later, Sir John would find himself in the papers again – this time because he was about to embark on the most perilous adventure of a lifetime; an expedition to chart the Northwest Passage through the frozen waters of the Arctic Ocean. The journey would involve traversing a route of over 1000 miles, from the Atlantic to the Pacific, negotiating notoriously difficult conditions, including sea ice and snow blizzards. It did come with a hefty pay packet, however, of £10,000 (close to one million today) to be distributed by Sir John amongst himself and his officers and crew. More importantly though, especially as far as Jane was concerned, Sir John would be hailed as the Discoverer of the Northwest passage if he were successful. At fifty-nine, Sir John was approaching the winter of his life, but with the formidable Jane to pull strings in all the right places, the position was secured for him.

Surprisingly, Jane would not be accompanying him - a decision which saved her life. But the reality was, there was never any question of her being there. The Arctic was a step too far, even for Jane. She was content to be an armchair explorer instead, tracking her husband's progress from afar via coordinates forwarded to the Admiralty in two-hundred tin cylinders that would be intermittently thrown overboard in the hope that they would be picked up.

'On Monday H.M.S. sloops *Erebus* and *Terror* left Greenhithe, on their attempt 'to penetrate the icy vastness of the north, and to circumnavigate America,'"[56] the *Illustrated London News* reported on 24 May 1845. The article was accompanied by a flattering illustration of Sir John himself – 'a portrait of the gallant Commander of the expedition,'[57] the paper announced with something of a flourish.

Jane was conspicuous by her absence at Sir John's departure. She had said goodbye the previous evening but chose not to go dockside the morning of his sailing, knowing that the press would give short shrift to her presence. The expedition was expected to last three years, which allowed Jane three years of almost uninterrupted travelling.

She went to France first, then Madeira, then travelled through the West Indies to the southern United States. She also climbed another mountain – this time the 6,300-foot Mount Washington in the White Mountains of New Hampshire, the highest peak in the northeastern United States. She was two hundred years too late to be crowned the first European to conquer it, however. That accolade had already been secured by Irishman Darby Field in 1642.

A report in the *Morning Herald* in November 1846 regarding her husband's progress suggested that some of the Inuit people in the Arctic had heard guns firing as if in celebration the previous autumn (of 1845) but Jane reasoned that because the two ships had not emerged in the Beaufort Sea, it was unlikely the Northwest passage had been conquered. Apart from this incident, the ships had not been seen since late July 1845, when they were encountered by the whalers *Prince of Wales* and *Enterprise* in Baffin Bay where they were waiting for the optimum conditions to enter the Lancaster Sound.

What happened next on the expedition would only be pieced together over the next one-hundred-and-fifty years.

By December 1846, no one had heard from either the *Erebus* or the *Terror* for over a year, and by 1848, public concern had grown sufficiently to set the wheels of bureaucracy in motion and a search commenced. Despite three separate searches – two travelling by sea and one overland - no trace of Franklin's expedition party was found. Even as late as 1850, false reports were circulating as to the fate of Sir John and his doomed expedition. 'An account just received from Liverpool,' wrote the *Nottinghamshire Guardian*, 'states that Sir John Franklin has been discovered by an English expedition on the Atlantic passage, in Prince Regent's Inlet, where he had been frozen in nearly four years. The expedition referred to had been able, it is said, to penetrate the icy barrier, and rescue the unfortunate adventurers.'[58]

For Jane, the lack of concrete information on the fate of the expedition, along with the wait for news, would have been agonizing. A letter she penned in 1853, eight years after Sir John left on his expedition, and what would transpire to be many years after his death, shows her complete and unswerving devotion to the search for him, and for Sir John himself: 'You must always have felt that I would never rest till I had more tidings of you....It is my heart's sole thought; the one and only object and occupation of all my faculties and energies. My own dear husband it is for you I live.'[59] The Admiralty declared the crew of Franklin's expedition as 'deceased in service' on 31 March 1854, a year to the day after Jane penned her pledge to Sir John.

Between 1850 and 1857, Jane financially supported the fitting out of five separate search ships, but only one search determined the awful truth. The *Fox* expedition, under Leo McClintock, which sailed from Aberdeen on 2 July 1857 was finally able to piece together the full horror of what happened. On 5 May 1859, a search party sent out from the *Fox* braved the freezing tundra of King William Island and discovered an official Admiralty document in a cairn. This document, now known as the Victory Point note, contained two messages from the Franklin expedition, written a year apart. The first dated 27 May 1847, said that both the *Erebus* and *Terror* had wintered in the ice off the northwest coast of King William Island, and previously at Beechey Island. It read: 'Sir John Franklin commanding the Expedition. All well.' The second message was written around the edge of the first and dated 25 April 1848. It reported that the *Erebus* and *Terror* had spent 18 months trapped in ice, with the crew finally abandoning the ships on 22 April. Twenty-four of the crew had perished, including Sir John Franklin who had outlived the first message by only two weeks. He had died on 11 June 1847. The note also stated that the remaining 105 survivors, who were now severely weakened by starvation, scurvy and exhaustion planned to head south the following day to the Back's Fish River, led by Captain F. R. M. Crozier.

The note was given further credence by the discovery of a skeleton on the southern coast of King William Island, which was still clothed and in possession of a seaman's certificate. Further remains were found inside a lifeboat, along with abandoned equipment from the Franklin expedition, including boots, handkerchiefs and hair combs. Among the many books that were found with the two bodies was a copy of *The Vicar of Wakefield* by Oliver Goldsmith.

Over the next ten years, various makeshift camps, shallow graves containing skeletal remains, and fragments of clothing were gradually uncovered in locations across King William Island by further expeditions.

Jane was partaking in one of her favourite pastimes when the news of the Victory Point note reached her, having just climbed the Pic du Midi de Bigorre in the Pyrenees. She had left England in 1858 to explore Greece, North Africa, the Middle East and the Crimea with her faithful sidekick Sophia Cracroft - Sir John's niece and Jane's constant companion since their time in Van Diemen's Land. On their travels they managed several coups. Firstly, they managed to secure a meeting with the Queen of Greece and talked her into lending them a Greek navy warship, and secondly Jane became the first British woman, and one of the first civilians of either sex, to visit the sites of the Crimean War, including the ruined Sevastopol where they witnessed the devastation of war, written on the landscape; the remaining shelled buildings standing 'white and sharp against the clear sky...'[60]

After a winter sojourn back in England, Jane and Sophia returned to the continent in 1859 – this time southwest France and the Pyrenees, which is where she received news of Sir John's death from Leo McClintock – the final, unequivocal closure she so desperately needed.

In 1860, Jane was awarded the Founder's Gold Medal by the Royal Geographical Society (RGS) – the first woman to be awarded the accolade. She may not have travelled to the Arctic region herself, but her sponsorship of the search missions involved in looking for Sir John allowed more of the area to be discovered and mapped. Quite simply, without her meddling and stubborn refusal to stop searching for her husband, vast swathes of the Arctic would have remained undiscovered for many more years – perhaps decades. Her contribution to the world's knowledge of the Arctic at that time was unquestionable, and to be recognized by the RGS was an astonishing achievement for a woman. The most significant element of the award, for Jane anyway, was the citation. It was read aloud at the award presentation, which Jane herself did not attend, either because of a bad leg or because, like many learned societies at that time, inclusivity didn't rank particularly high on the RGS agenda:

> Desirous of commemorating in an especial manner the Arctic researches of our associate, the late Sir John Franklin, and of testifying to the fact that his expedition was the first to discover a North-West Passage, the council of the Royal Geographical Society has awarded the founder's gold medal to his widow, Lady Franklin (cheers), in token of their admiration of her noble and self-sacrificing perseverance in sending out at her own cost several searching expeditions, until at length, the fate of her husband has been finally ascertained (hear, hear)[61]

To which Jane replied in writing that the award testified 'to the life-long services of my husband in the cause of geographical research, and especially of the crowning discovery of the North-West Passage by himself and his companions, which cost them their lives.'[62]

The sixty-eight-year-old Jane was awarded her medal in true ladylike style - at home, reclining on the sofa.

Jane spent the early 1860's on the move again with Sophia Cracroft in tow, sailing first to North America and visiting New York City – then just an up-and-coming hub of commerce – before travelling north to Toronto and then back to New York via Niagara Falls. She then got on a ship and sailed to Brazil, visiting Rio De Janeiro and the 'Magnanimous' Emperor Pedro II, who was more than a match for her intellectually, although Sophia recorded that the royal interview lacked ceremony. Patagonia was next on the list, followed by Chile, Panama, Acapulco, San Francisco and then Vancouver Island, where they disembarked at Victoria. Jane was almost seventy, but that didn't stop her getting in a canoe – not once, but twice during her trip. From Vancouver Island, they went south again to Oregon, then California before setting sail on the *Yankee* for a two-week return trip to the Sandwich Islands, now known as Hawaii. After taking in the local landmarks, including the monument to Captain James Cook, Jane and Sophie sailed back to San Francisco, learning on their arrival that the American Civil War had broken out. The original journey back to England involved travelling first to New York, but with cross-country travel now out of the question, Jane decided to sail to Japan instead – an unexpected but welcome addition to their travels. The two ladies visited Yokohama, Yedo and Nagasaki before sailing on to Shanghai, Hong Kong and the British colony of Singapore. India was next so they crossed the Bay of Bengal, dined in Calcutta with Lord Elgin and then swung around the Cape of Good Hope to arrive back in England in July 1862.

Jane managed to stay put in England for two years before they set off again. This time they went to Spain, where she rode a donkey 4000ft to the peak of the Monserrat mountains. Then they sojourned in Majorca, long before it became the mass tourist destination it is today before journeying home via Madrid.

And this would be the story of the final ten years of Jane's life – short periods in England, followed by extended periods of exploration. A last hurrah to the pursuit she had held most dear. Egypt, India, Ceylon and Alaska were all traversed and conquered in the winter of Jane's life, as well as return trips to North America and some countries closer to home such as Switzerland, Italy, France and Germany.

During this period, Jane also kept her mind firmly on matters concerning her husband and his legacy. As far as she was concerned, Sir John deserved to have his memory venerated with at least one memorial, if not several – preferably in central London. This was critical to Jane's campaign to ensure her husband was remembered forevermore as a person of note in Arctic exploration, not just in the Admiralty, but by the entire country.

Jane would succeed in this, as she did in most things – through sheer force of will and perseverance. She had conquered real mountains, so the metaphorical ones were of little consequence. Needless to say, she got her statues – although she had to concede on one location. Jane wanted Trafalgar Square (if it was good enough for Lord Nelson, it was good enough for Sir John) but instead she got Waterloo Place.

Sir John's statue was unveiled in 1866 and deemed a success; *Saunders's Newsletter* called it a 'beautiful monument,'[63] and Franklin's likeness was described as, 'characteristic and excellent'[64] by Jane. The plinth of the Franklin memorial depicts what his funeral may have looked like – although of course there is no record of what exactly took place. Two casts were made; the other was to be shipped to Van Diemen's Land (now Tasmania) and erected in Hobart. Five years earlier, the town of Spilsbury, Sir John's birthplace, had already erected a monument to its famous son. Jane wrote the inscription herself, and it reads, predictably, 'Sir John Franklin / Discoverer of the Northwest Passage.' Yet both memorials were eclipsed by his magnificent white marble bust in that most hallowed and prestigious of locations – Westminster Abbey. Matthew Noble's resplendent sculpture, which was erected in 1866, shows Franklin gazing in the direction of lands afar, a look of dogged determination etched into his heavy brows. The verse was penned by none other than the doyenne of Victorian poets Alfred, Lord Tennyson, and read: 'NOT HERE: THE WHITE NORTH HAS THY BONES; AND THOU, HEROIC SAILOR-SOUL, ART PASSING ON THINE HAPPIER VOYAGE NOW TOWARD NO EARTHLY POLE.'

As for Jane, there was no memorial forthcoming when she herself died in 1875 at the age of eighty-three. It was typical of her to thrust her husband forward as readily as she thrust her opinions on everyone. But she did lend her name to several natural features and locations – Lady Franklin Bay on Ellesmere Island, Canada's northernmost island; Lady Franklin Rock in Yosemite National Park, California, which Jane visited in 1861; Mount Lady Jane Franklin, a hill in Northern Victoria; and finally, Lady Franklin Rock, an island located in the Fraser River, which is described by the Encyclopedia of British Columbia as 'a large obstacle in the Fraser Canyon,' which seems about as fitting a tribute to Jane Franklin as it is possible to get. Naturally, she also had several ships named after her - although one of these, rather ominously, sank in 1856, and another caught fire in mysterious circumstances in 1861.

There is a sense with Jane that her pleasure in life came from living out her own dreams and ambitions vicariously through Sir John. She was intelligent enough to recognize the limitations of her sex, and whilst she consistently struck out on her own with her own agenda, her attachment to Sir John ensured that she was accepted into the exclusive, and hitherto male-dominated 'elite explorer club.' Not only that, her extensive grasp of geography, topography, history, politics and culture also meant she was taken seriously, rather than perceived as just another 'lady-traveller.'

'Jane, Lady Franklin, widow of the great Arctic explorer, Sir John Franklin, K.C.H., died at her residence in Phillimore Gardens on the 18th,'[65] reported the *Illustrated London News*, in July 1875. She was eighty-three, and despite her rapidly declining health had spent her final days 'employed in the prosecution of the pious task to which she had devoted the later years of her life, by assisting her friend Sir Allen Young to fit out the *Pandora*, which has just sailed in search of any additional records that may yet be recoverable of Sir John Franklin and his companions.'[66]

Which is exactly what Jane would have wanted.

Rather than hog the limelight with her own memorial, Jane was content to share one with her husband. As a Christian – this would be the most important journey of her life, for it was the one that would at last reunite her with her husband; a sentiment echoed in the final, touching inscription: 'AFTER LONG WAITING, AND SENDING

MANY IN SEARCH OF HIM, HERSELF DEPARTED, TO SEEK AND FIND HIM IN THE REALMS OF LIGHT...'

* * * * *

Some women travelled with romance at the bottom of their list of reasons to explore. Embarking on an adventure of her own design was scandalous enough without engaging in a romantic liaison along the way. And in many cases logistically impossible too, because unless her admirer was prepared to relinquish his own needs and responsibilities and travel along with her, the relationship was doomed from the start.

With the odds stacked against them, those explorers who did find themselves falling in love were surprised by the turn of events. None more so than Isabella Bird, who frequently made companions of the men she met on her travels - even if it only meant she had someone to hold onto whilst she peered over the edge of a volcano. Isabella thought nothing of hunkering down with the natives and frequently shared tents and sleeping arrangements with the opposite sex. When a gentleman traveller, known only as Mr Green, promised Isabella fireworks (of the fiery ember kind - as far as she was concerned anyway) at the top of Hawaii's largest volcano, Mauna Loa, Isabella was quick to agree. In fact, she barely batted an eyelid when he suggested they share a tiny tent, along with their all-male group of native guides. Mr Green appears to have been most useful – sourcing mules, packhorses, and numerous guides; even providing Isabella with a 'French soldier's overcoat, with cape and hood,'[67] because she was getting cold. In fact, he seems to excel himself in all areas and be quite the gentleman – apart from the fact he forgot to pack the tea for their ascent of the volcano ('[I] then felt how impossible it was to exert myself in the rarified air, or even to upbraid Mr Green for having forgotten the tea, of which I had reminded him as often as was consistent with politeness!'[68]) This was clearly a deal-breaker for Isabella but apart from this mishap (his only) he appears to have been the perfect travelling companion. Lillias Campbell Davidson would most certainly approve, even if she didn't condone the sharing of the tent, or the forgetting of the tea. Especially the forgetting of the tea: 'In travelling it is as well to take with one one's own tea,'[69] she writes.

So, when Isabella met the wild and wayward Jim Nugent during her stay at Estes Park in the Rocky Mountains in 1873, he seemed an unlikely match. She was a formidable, forty-something spinster with a severe aversion to 'masculinity,' he was a gun-toting, one-eyed Irishman with a penchant for heavy-drinking and violent moods. It had disaster written all over it.

After their first meeting, Isabella described him as, 'Broad and thickset…about the middle height, with an old cap on his head, and wearing a grey hunting suit much the worse for wear (almost falling to pieces, in fact), a diggers scarf knotted around his waist, a knife in his belt, and a 'bosom friend', a revolver sticking out of the breast pocket of his coat.'[70] She was intrigued, revealing that she found his face 'remarkable', and believed that he must have been 'strikingly handsome' in his youth. She continues,

> He has large grey-blue eyes, deeply set, with well-marked eyebrows, a handsome aquiline nose, and a very handsome mouth. His face was smooth-shaven except for a dense moustache and imperial. Tawny hair, in thin uncared-for curls, fell from under his hunter's cap and over his collar. One eye was entirely gone, and the loss made one side of the face repulsive, while the other might have been modelled in marble.[71]

Lust at first sight? Perhaps. "Desperado' was written in large letters all over him,' Isabella wrote. 'I almost repented of having sought his acquaintance.'[72] Note – *almost*. Well…at least she knew what she was getting into.

Their first conversation started well. 'He told me that the loss of his eye,' wrote Isabella, 'was owing to a recent encounter with a grizzly bear, which, after giving him a death hug, tearing him all over, breaking his arm and scratching out his eye, had left him for dead.'[73] Clearly Mountain Jim was a bit out of practice when it came to making appropriate small-talk with the opposite sex. But then Isabella wasn't like most women, and if anyone was to be impressed by his tale of derring-do, it was her. 'As he spoke I forgot both his reputation and appearance,' she wrote, 'for his manner was that of a chivalrous gentleman, his accent refined, and his language easy and elegant.' Her interest was piqued, and her desire to climb the vast Colorado mountain ranges provided the perfect excuse to get to know America's answer to Emily Bronte's Heathcliff. Jim was to be her guide, along with 'two youths.'

Isabella's account, which she published in *A Lady's Life in the Rocky Mountains* (1879) is of course one-sided. We glean nothing of how Jim Nugent initially felt about the 'solid and substantial little person,' he was to become acquainted with – not until much later anyway. But for all his 'daring exploits,' and 'ugly fits,' Mr Nugent proved himself to be nothing but noble in the company of Miss Bird. When she had an attack of giddiness and fatigue ascending the 14,000ft Long's Peak mountain, Jim was there to catch her and 'drag' her along. He also went all out to show Isabella his cultured side, performing both 'Indian stories' and 'a very clever poem of his own composition' once they reached their camp for the night. He didn't neglect to show his vulnerable side either. The two 'friends' sat up talking for most of the night, with Jim regaling Isabella of 'stories of his early youth' and confessing 'a great sorrow which had led him to embark on a lawless and desperate life,' which was also accompanied by a trembling voice and tears which 'rolled down his cheek.' Jim might have been making every effort to impress, but Isabella was no fool. 'Was it semi-conscious acting,' she wondered, 'or was his dark soul really stirred to its depths by the silence, the beauty, and the memories of youth?' It's clear she hoped it was the latter.

We get the sense with Isabella as she gazes longingly at Rocky Jim - his face no doubt looking even more handsome in the flattering light of the fireside – that he is just out of reach, either because of her own sense of propriety or because of their obvious, and insurmountable class difference. Jim is just a little too far over the wrong side of the tracks to be considered a serious suitor – but it's comforting to know that even in the nineteenth-century, sensible, intelligent women such as Isabella Bird could just as easily have their hearts set aflame by the attentions of a notorious 'bad-boy' as her twenty-first century descendants.

In some respects, Isabella's account of her fledgling intimacy with Rocky Jim, and the romantic narratives of explorers such as Margaret Fountaine could have been written in the modern era. Their unconventionality allowed them to take a novel and open-minded approach to speaking plainly about their feelings – with none of the public reticence that stifled many women of the time. That is not to say other women were not expressive in private – quite the contrary - but such public declarations of emotion were unusual, and to broadcast it through a published account was highly subversive.

But at what price? Contemporary reactions say much about how this break from conventionality was received. On the release of Isabella's travelogue, *A Lady's Life in the Rocky Mountains* in 1879, the press responded by practically snorting with derision at her life-choices:

> One expects to find heroes braving all manner of dangers but when a heroine like Miss Isabella Bird goes and lives among the most desperate characters in the Rocky Mountains for months at a time disguised as a man, and having friendship with the most dreadful ruffians, one wonders what is next to come. This lady explains that she travelled in this way for her health. We may congratulate her upon the possession of a sound nervous system, whatever other infirmity she may have laboured under.[74]

Thus, to make her 'friendship with the most dreadful ruffians' acceptable, it became necessary to obscure the truth by suggesting that she only did so because she was 'disguised as a man.' That a man and woman from entirely divergent backgrounds and opposite ends of the social scale could have an intimate relationship – platonic or otherwise – and divulge the fact to the rest of the world was anathema to nineteenth-century sensibilities. Like many of her contemporaries, Isabella didn't go out of her way to scandalize – it was just an unfortunate by-product of her forward-thinking approach to life. Victorian society just wasn't quite ready to accept anything or anyone who upset the status quo.

Like all the best love stories, Isabella and Mountain Jim were destined for a tragic end. Jim eventually declared his admiration for Isabella, but his propensity to be honest with her in all things – including his dark past – was his undoing. That, and the enormous amount of whisky he consumed daily. Isabella later revealed that the things Jim had revealed about himself the afternoon he had pledged his devotion had 'come between me and the sunshine, sometimes, and I wake at night to think of them.'[75] She also tried tough love: 'for the last time I urged upon him the necessity of a reformation in his life beginning with the giving up of whisky, going so far as to tell him that I despised a man of his intellect for being a slave to such a vice. 'Too late! Too late! 'he always answered, 'for such a change.' Ay, *too late*.' Unfortunately, Isabella's admiration was rapidly turning to pity. 'He shed tears quietly. 'It might have been once,' he said. Ay, *might* have been. He has excellent sense for everyone but himself…'[76] was Isabella's final assessment of the man she almost loved. She was also frustrated, because she couldn't help but wonder what he could have become had he not chosen the path of ruin. She was also annoyed that he didn't have the strength of character to overcome his challenges – something a lady explorer, especially the indomitable Isabella, would never countenance. His apathetic acceptance of his situation just wasn't good enough.

Following the dénouement of the Isabella-Jim romantic tale, Isabella set her sights on adventures anew, journeying to Persia, Japan and China. Sadly, Mountain Jim came to a sticky end five months after Isabella left the mountains. He was killed by a long-standing nemesis, Griff Evans, who shot him from his doorstep whilst Nugent rode past his cabin. Several different versions of the tragedy and how it happened eventually reached Isabella. But all had the same, sorry outcome. 'The story of the previous weeks is dark, sad, and evil,' she wrote. 'Of the five differing versions which have been written to me of the act itself and its immediate causes, it is best to give none. The tragedy is too painful to dwell upon.'[77]

Jim died how he lived, which was thoroughly in keeping with his tragic hero persona. As for Isabella, she did eventually marry… this time to a doctor – a sensible decision but also a remarkably conventional one for a woman like her. With her sanity still intact, and Victorian rationality finally kicking in, Dr Bishop was the natural choice. Even the grand dame of exploration had her limits.

Part 3

For science and antiquities

'For long excursions into the desert Camels will be required. The ordinary baggage-camel is very heavy and rough in his paces, and it requires considerable experience in camel riding before the *Hegin* or trotting camel can be mounted with any comfort. The paces of a quiet, smooth-walking camel are, however, by no means unpleasant.'[1]

Mary Brodrick, 1900

Mary Brodrick knew the best way to traverse the Egyptian desert better than anyone. As one of the first female archaeologists to excavate in Egypt, her experience and expertise were renowned, despite being told that as a woman she had no business studying Egyptology. In fact, the challenges brought about by her sex began before she'd even picked up a pen.

Born in London in 1858 to solicitor Thomas Brodrick and his wife Mary, the young Mary's interest in ancient Egypt began in childhood when she was gifted a book about Egyptian history. Against the backdrop of a booming tourism industry, courtesy of Thomas Cook and his package tours of Egypt and the Nile, and the earlier works of Sophia Lane Poole (*The Englishwoman in Egypt: Letters from Cairo* (1842)) and Amelia Edwards (*A Thousand Miles Up the Nile* (1877), and co-founder of the Egypt Exploration Fund), Mary was keen to see the British-occupied Egypt for herself and embarked on her first journey up the Nile at the earliest opportunity. With her interest piqued - and no doubt inspired by the gravitas of her ancient surroundings - she set about improving her knowledge of the region on her return, but there was a major sticking point. There were no suitable courses for her chosen subject in London, and certainly none that would accept a female student. Undeterred, Mary looked further afield – finally setting her sights on the Sorbonne in Paris, under Professor Gaston Camille Charles Maspero, a world-renowned Egyptologist who had served as the director-general of the Egyptian antiquities service. Maspero was less than impressed by the idea of a woman sitting in his lectures, reputedly declaring, 'But we don't take little girls here.'[2] Other lecturers were equally unenthusiastic. At Maspero's suggestion, Mary approached Joseph Ernest Renan, expert of Semitic language and civilizations at the Collège de France. His response was, 'I have never taught a woman in my life, and I never will!'[3]

Maspero did at least show a flicker of sympathy for her plight. He raised her case with the council of the Sorbonne, and as a result of his intervention (and the college rules that had overlooked any clauses relating to the sex of their students) she was finally allowed to attend lectures at both the Sorbonne and the Collège de France. That was only half the battle won, however. She had to win over the respect of her contemporaries for her to be taken seriously – amongst them some of the greatest names in the field; Georges Legrain and Georges Aaron Bénédite, amongst others. First impressions were not good on either side, with Mary writing later in her personal papers that: 'The students gave me a bad time; they were rough and rude and they smelt,'[4] although eventually they did become 'great friends,'[5] despite pouring ink down her back on at least one occasion.

On Mary's return to London, she began lecturing at the British Museum on Saturdays and became an 'occasional student' of the institution under the tutelage of Egyptian Antiquities Keeper, Philip LePage Renouf. The two had a close working relationship. Mary recalled later in life that on one occasion Renouf had given her a hieroglyphic inscription to read: 'I took it back humiliated and told Renouf that I could make nothing of it. The old man looked up with a charming smile and said: "Neither can I," and added that he had given it to me to see if I had any grit.'[6]

By 1888, she had enrolled at the only women's university residence in London, College Hall, studying Egyptology under Dr Stuart Poole (the son of Sophia Lane Poole), and continued to lecture – not just to display her knowledge of ancient Egypt, but to champion women's education. 'Miss Brodrick is a very gifted lecturer,' wrote the *Eastern Daily Press*. 'In an hour and a half she put before her audience, which ought to have been larger, in a lucid and graphic manner a vast amount of information about ancient Egypt.'[7] Mary also knew the benefits of engaging her audience with information they could relate to – the culture and customs of the people, particularly the experience of women. 'The dignity and position of women was a remarkable fact,' the paper related after her lecture. 'Education was free and compulsory, both for boys and girls. From the age of four to that of fourteen, boys and girls, went to temple schools, those of princes and peasants alike, there to learn together.'[8] The visibility of notable female scholars such as Mary and her contemporaries Helen Tirard, Jane Harrison and Anna Morton, helped to raise the profile of women as archaeologists, and the idea that they too could become experts in their field alongside their male contemporaries.

But what Mary really loved best was touring and working in the field. From 1894-1896, Mary held the Pfieffer fellowship, awarded by the Council of College Hall, which allowed her (like its namesake) to travel in Italy and Greece, and more extensively in Egypt, the region she loved best. 'The Nile is emphatically the river of the Old World,' she wrote in *A Handbook for Travellers in Lower and Upper Egypt*. 'No river, in either hemisphere, can in any degree equal it in historical and geographical interest. By the ancient Egyptians it was honoured as a divinity to whom their land was indebted for its very existence, Egypt being most truly, as Herodotos puts it, "the gift of the Nile." Its connection with the most important events of ancient history, and the stupendous monuments which still bear witness to its former wealth and civilization, render it an object of the greatest interest to the archaeologist and the student of history; while the discovery of its source has been a problem which down to the present day has never ceased to excite the curiosity and stimulate the zeal of geographers and travellers.'[9]

Egypt had captured the public imagination and Mary was instrumental in opening up the region to tourists and travellers. In 1893, Mary submitted a manuscript to the publishers John Murray for a short handbook full of useful tips for those planning a trip to Egypt. Drawing on her extensive knowledge of the country, she was able to offer not only logistical information such as what clothing to pack, where to find the best hotels and what the local currency looked like but suggested routes that a traveller could take in order to soak up the very best of ancient Egypt – with a handy historical narrative to follow so that everyone could understand and appreciate the monuments and landscape. Mary took this one step further in 1897, becoming an in-person tour guide alongside her contemporary Anna Morton. The tour, titled 'A Winter on the Nile in a Dahabiyeh,' was exclusively for women and Brodrick was keen to ensure her full credentials were recognized on the promotional material, where she appeared as 'Miss Brodrick, Ph.D' rather than the customary, 'M. Brodrick.' She followed this up in 1903 with Murray's *Handbook for Travellers in Syria and Palestine.*

'Miss Mary Brodrick, Ph. D., who is delivering a series of lectures on ancient Egypt in the British Museum, knows that country better than any living woman,' reported the *Penrith Observer* in 1898. 'Her acquaintance with it dates back to the winter of 1888, and she has lived there more or less ever since. Her indefatigable energy in archeological research has greatly increased our knowledge of the history, region, and architecture of the ancient inhabitants of that country.'[10]

After Mary's fellowship expired, she was also commissioned to translate several key Egyptology texts under her former tutor, and director-general of Egyptian antiquities, Professor Maspero. Her translations of Auguste Mariette's 1867 work *Aperçu de l'histoire ancienne egypte* (re-titled *Outlines of Ancient Egyptian History*) and Heinrich Brugsch's *Geschichte Aegypten's unter den Pharaonen* (re-titled *Egypt under the Pharaohs: a History Derived Entirely from the Monuments*) opened up the subject to an international audience during the boom years of Egyptology and tourist travel to North Africa. Mary ensured her writing was more accessible to the amateur scholar and historian, whilst also reflecting the most recent discoveries in Egyptological science and the most up-to-date excavations and any new discoveries.

By 1906, the Daily Mail were hailing Mary as 'perhaps the greatest lady Egyptologist of the day'[11] at a time when interest in Egypt and ancient civilization showed no sign of diminishing. With the opening of the Cairo Museum of Antiquities in 1902 and new discoveries in the Valley of the Kings, it was not only an intense period of discovery in Egypt, but also heralded a dawning recognition that the security of ancient sites could be easily compromised. Broderick's winters in Egypt on her Nile *dahabiyeh* (a passenger boat – flying her own embroidered pennant no less) gave her first-hand experience of the conflicting notions of making ancient Egypt accessible while at the same time safeguarding its preservation for future generations. In her posthumously published personal papers, Brodrick recalled an incident at Luxor, when grave-robbers smuggled a mummy onto her dahabiyeh one night (against her wishes) so that Professor Maspero would not find it.[12] It wasn't the first time she'd be brought face to face with the destruction of an ancient site. As early as 1890, Brodrick had written to the *Times* reporting that she had 'caught two Egyptian soldiers in the act of cutting their names on the entrance of Aboo [sic] Simbel.'[13] Such observations led her to become involved with the Society for the Preservation of the Monuments of Ancient Egypt, as well as her continued committee membership of the Egypt Exploration Society.

Mary's legacy was the ability to bring the colour and vibrancy of ancient Egypt to life in the dusty town halls and lecture theatres of an (often) chilly and grey British Isles. She brought a little bit of Egypt home with her every time she returned to Britain – not in a physical sense; but with her skill in conveying a common history that turn-of-the-century Britons could relate to, despite being separated by thousands of years. It was Egypt as she saw it in her mind's eye – which would always be that of an Egyptologist - but communicated in a way she knew would be easily digestible. 'For upwards of an hour Miss Brodrick, who dispensed with a chairman, held the keenest attention of her audience, addressing her remarks in pleasant, conversational style,'[14] wrote the *Eastern Daily Press* in 1902 of her lecture, 'A Day in the Life of an Ancient Egyptian.' One of her final lectures was given following one of the greatest archaeological discoveries of the era – the unearthing of the tomb of Tutankhamun in November 1922 – which not only ignited a new period of Egyptomania but introduced a sense of ghoulishness and mystery in the public imagination.

Mary Brodrick died in 1933 – the same year that Boris Karloff portrayed an Egyptologist who rises from the dead to haunt everyone who betrays him in horror film *The Ghoul* (1933). Looking as 'unpleasant as ever,'[15] according to one review in the *Exeter and Plymouth Gazette,* what Mary would have made of the famous actor's interpretation of an Egyptologist we will never know, but to those who knew her she was far from the 'revolting' gothic character offered up by Boris Karloff. 'She was among the most distinguished that belonged to us. She was one of the best-known pioneer women Egyptologists,'[16] was Dr Louisa Macdonald's tribute at an address in November 1933. The *Times* was more intimate: 'one of the kindest, wisest, most generous and true of friends,' was their appraisal of Mary's character. She pioneered both her field of expertise and her sex at a time when women's higher education was in its infancy – and she did it with all the aplomb of her more-famous male contemporaries to boot.

* * * * *

'MISS MARIANNE NORTH'S SKETCHES. Those visiting The Gallery, 9, Conduit street, Regent-street, containing Miss North's 500 studies in oil, will probably be disappointed if they expect to find works which will bear strict examination as to their technically artistic merits, for many of the sketches are crude in colour, not particularly well-drawn, and oblivious of the Rules of Art affecting composition and light and shade.'[17]

That those same 'crude in colour' paintings would three years later find themselves a permanent home at the Royal Botanic Gardens at Kew, to delight and inform generations of visitors, probably came as a surprise to the journalist of the *Daily Chronicle.* It was no surprise to Marianne herself, of course. She knew her talent and skill as an artist went beyond mere technical merit. Just as Mary Brodrick had opened up ancient Egypt to the masses, Marianne North's collection of botanical paintings brought a taste of the exotic to the general public. From the heady gardens of the Taj Mahal to the steep, stony slopes of the Cordilleras of Chili, Marianne was able to communicate the diversity of natural life in the obscurest corners of the globe via her paintings. And just as her paintings defied the rules of botanical and scientific art, so Marianne defied the rules of convention in order to amass her remarkable collection.

Fortunately, the *Chronicle* and their scathing assessment of Marianne's work were in the minority. *The Graphic* was quick to recognize Marianne's talents, writing that her work was one of the most 'remarkable, instructive, and interesting Exhibitions ever brought within the walls of the South Kensington Museum.'[18] The critic continued,

> When to these facts it is added that Miss North displays in her work not only the freedom of hand, the purity and brilliancy of colour, and the accurate draughtsmanship of a consummate artist, but that she is also an excellent botanist and a keen observer of Nature, it will be seen that her collection is one of very unusual value, and that the Government has done well in thus borrowing and exhibiting it for the gratification of the public.[19]

As the eldest daughter of a well-to-do land-owning gentleman, Marianne had picked up a paint brush in her adolescence, much like other young ladies of the time. But where her contemporaries were happy to paint the landscapes and domestic scenes they lived in, Marianne set her sights on travelling a bit further afield to perfect her talents. Born in 1830 in Hastings, Sussex and with an aristocratic pedigree stretching back to James II and the Reformation, Marianne's childhood years were spent wintering in Hastings, before travelling to London for the spring and then dividing the summer between her half-sister's hall in Lancashire and a farmhouse in Rougham, Norfolk. It was a happy time; made happier by the close relationship she enjoyed with her father, recalling in her aptly named autobiography *Recollections of a Happy Life* (1894) that, 'He was from first to last the one idol and friend of my life, and apart from him I had little pleasure and no secrets.'[20] Following a trip to Kew gardens, she was gifted an *Amherstia nobilis* (Pride of Burma) by Sir William Hooker, director of the gardens, which inspired her desire to see the tropics.

After the death of her mother in 1855, Marianne, her sister Catherine, and their father spent the summers abroad, travelling extensively on the continent (or 'wandering abroad,' as Marianne called it) touring Switzerland and the Alps, Hungary, before making their way south to Constantinople and Athens.

A stroke of bad luck in 1865 (her father lost his Parliamentary seat) meant that Marianne was able to start planning a much-longed for trip to Egypt, and following a very troublesome journey during which they were forced to quarantine at several locations – the worst in Marianne's opinion being Alexandria, a 'nasty, mongrel, mosquito-ish place, and we got out of it as fast as we could'[21] – they eventually arrived at Cairo. Their journey up the Nile was so slow due to a lack of wind and too much mud that Marianne had plenty of opportunities to go on shore and draw her ancient surroundings, as well as the river wildlife which she found 'in abundance to entertain us on shore or afloat.'[22]

In 1869, Marianne was left devastated by the death of her father – a painful recollection that she skips over in her autobiography, merely noting the haste of her beloved father's demise and his final heartbreaking words: 'Come and give me a kiss, Pop, I am only going to sleep.[23] 'He never woke again,' Marianne wrote, 'and left me indeed alone.'[24]

Tragic though the circumstances were, Marianne had fulfilled the duty expected of her by society – and she had done so willingly. Her father was not only her travelling companion but her one constant – a lodestar that had guided her through the first forty years of her life. 'He had been my one friend and companion, and now I had to learn to live without him, and to fill up my life with other interests as best I might.'[25]

Just as Margaret Fountaine had used butterflies to distract her from her romantic misadventures, Marianne found a means to escape her heartbreak by devoting her life to painting. She would travel less conspicuously than some of her lone explorer sisters, deciding very quickly that (initially anyway) she would take her old servant Elizabeth with her as a companion and chaperone. Marianne was nothing if not respectable, and she intended to follow convention to the letter – rather than sully the good name of her esteemed ancestors. This did not automatically confer a mutual friendship, however, and Marianne soon became exasperated by Elizabeth's company, complaining in her diary that she found their constant companionship 'very wearisome'. Little wonder that on most of her later trips she decided to travel alone.

In 1871 she was ready to begin her quest, travelling initially to Canada and the United States, and then on to Jamaica and Brazil. Marianne was overwhelmed by the American hospitality ('huckleberry puddings with cream were quite divine'[26]) but probably a little underwhelmed by the native species she was there to paint, although she found beauty in 'a most curious old cedar-tree, perfectly shaved at the top like an umbrella pine by the sea winds, with its branches matted and twisted in the most fantastical way underneath…'[27]

Canada was more promising, with the spectacular Niagara Falls giving Marianne hours of pleasure in its painting. Her sense of wonder is palpable in her description of the famous falls which, she recalled, 'tempted me much, standing close to its head, with the rapids like a sea behind, and the rainbow dipping into its deep emerald hollow; the tints were endless in their gradations, and delicious, but I got wet through in the mist.'[28]

It was all rather glorious, but Jamaica and the West Indies were her most highly anticipated landscape. Even a meeting with the United States President – at that time Ulysses S. Grant – couldn't compare to her excitement at arriving in the tropics, although the President and First Lady did themselves no favours by getting Marianne confused with the daughter of an eighteenth-century peer. Marianne wondered about the fuss that was made of her, and it became apparent after the meeting that, 'Mrs Grant [the First Lady] talked of me as the daughter of Lord North, the ex-Prime Minister of England. I always knew I was old, but was not prepared for that amount of antiquity.'[29]

When Marianne finally reached Jamaica ('In the West Indies at last!') it was as though her senses had been assaulted. She describes the 'richest foliage' of 'bananas, rose-apples (with their white tassel flowers and pretty pink young shoots and leaves), the gigantic bread-fruit trumpet-trees (with great white-lined leaves), star apples (with brown and gold plush lining to their shiny leaves)…' etc. etc. In what must have been a feast for a painter's eyes, Marianne spent from sunrise until noon painting, and then spent the rainy afternoons pottering and painting in the house.

The vivid colours of the tropics suited Marianne's medium of oil paints, which served to further intensify the vibrancy and brilliance of the plants. In this regard, she was something of a trendsetter, preferring to buck the usual trend for botanical watercolours for something punchier. Her trick of depicting a plant or flower in its natural habitat gave life and context to her work and allowed the viewer a glimpse into the landscapes she herself had visited. Mountains, temples, people, insects, birds and animals – all featured in Marianne's work – giving it an extra-dimension and no doubt raising a few eyebrows in the establishment in the process.

But Marianne's work went beyond mere artistry, she also contributed to botanical science. Her travels enabled her to discover plants that were entirely undocumented in Europe. In Borneo she painted a pitcher plant that grew in the limestone mountainous regions, which caused much excitement in the botanical fraternity. She eventually lent her name to it - *Nepenthes northiana* – which must have been a huge honour to the unassuming Marianne (and well-deserved too, given that she'd endured leeches and swamps in order to paint Borneo's finest specimens). Marianne also later had several other species named after her – *Northea seychellana, Areca northiana, Chassalia northiana, Crinum northianum* and *Kniphofia northiae* – in recognition of her discoveries.

Marianne spent the next fifteen years travelling. In fact, she circled the globe twice, in both directions, with short interludes in London during which she exhibited her latest paintings, planned her next overseas adventures and donated (and had built) an entire gallery at the Royal Botanical Gardens at Kew, in order to permanently house her now vast collection.

Her travels were well documented by the press, who were quick to inform their readers of her whereabouts. 'Miss Marianne North has sailed for the Cape to resume her task of painting the flora of all parts of the world,' announced the *Leeds Times* in 1882. 'After spending some months in South Africa, she proposes to visit Madagascar and the Seychelles Archipelago, both of which present rare and beautiful objects for the pencil. As before, she travels alone and unattended.'[30] She also reached the heady heights of being listed as a 'Distinguished Lady,' in the London Gossip column of the *Birmingham Post*. 'The ladies are decidedly coming to the front in most bold and gallant style,' reported the paper, complimenting her on her 'thirst for knowledge, not for the mere selfish purpose of enjoyment, but for the sake of utilising it for the benefit of others.'[31] The *Illustrated London News* regularly printed sketches of her, sitting at her easel, palette in hand – drawn from photographs that had been forwarded to them from abroad. Perhaps the most famous photographs of Marianne working in the field were those taken by the pioneer portrait photographer Julia Margaret Cameron at her home in Ceylon. It was an experience Marianne documented in her autobiography, describing how Mrs Cameron 'made up her mind at once she would photograph me.' Afterwards, Marianne's opined that the twelve plates she took were 'all in vain' and that the images had turned out 'perfectly uninteresting,' but she could at least take comfort in the knowledge that as one of Cameron's subjects she was in good company with the great and good of the era such as Charles Darwin and Alfred, Lord Tennyson.

Despite failing health Marianne's final journey took her to Chile, in search of the one tree she felt ought to be represented in oils in her gallery, but until that time was missing – the *Araucaria imbricans* or 'monkey-puzzle' tree. 'My chief object in coming to Chili [sic]', Marianne wrote in a letter to the press, 'was to see and paint the old forests of *Araucaria Imbricata*, known in England as the puzzle-monkey tree, rather unreasonably, as there are no monkeys here to puzzle; probably, they crossed the Cordilleras in disgust at the general prickliness of all plants here, especially the araucarias, and never came back again.'[32]

Marianne's understated humour and turn of phrase marked her out as an accomplished writer as well as a skilled artist, and her autobiography is a celebration of all that is good and beautiful in the world. From the loveliness of the full moon in Borneo; to the 'beautiful turquoise and lapis lazuli' of the Indian minarets; and the delicious fruit of the Seychelles - Marianne delights in all she has seen and invites the reader to delight in it too.

When her health began to fail during her final trips – exacerbated by nervous anxiety – Marianne reflected on her 'happy life.' But there was no place she'd rather be at the end than England: 'No life is so charming as a country one in England, and no flowers are sweeter or more lovely than the primroses, cowslips, bluebells, and violets which grow in abundance all round me here.'[33] She had found her perfect spot, and it was at her home in Alderley, Gloucestershire, where she had cultivated her own pastoral idyll, that she quietly died on 30 August 1890.

Part 4

For fame, fortune and newspapers

'Miss Londonderry is…only 23 years old…and in her riding trope presents a very attractive appearance… The young woman is a sort of riding advertising agency. She wears ribbons advertising various goods and will receive $400 for one firm's ad that graces her left breast. On her right bloomer leg she carries $100 worth of advertisements and she has just closed a contract to cover her left arm. She says her back is for rent yet and she hopes to get $300 for it. She must not beg a cent and makes enough money to pay for her board by selling little souvenir Brownie pins and other souvenirs of the trip.'

The Buffalo Express,
1 November 1894.

As Annie Londonderry sailed into Buffalo in 1894, a cool breeze blowing on her face and the autumn sun warming her back, the working class life she had been living in a tenement building in Boston's West End – the life she had cycled away from six months earlier - was the last thing on her mind. Instead, she had refashioned herself into a one-woman cycling extravaganza – a 'circumcycler' as one newspaper named her, in honour of her daring attempt to become the first woman to circle the globe on a push-bike. The idea was preposterous, given Annie's utter lack of preparation (she had never even sat on a bike let alone ridden one until the moment she set out) but Annie was nothing if not plucky – and pluck was something she had in abundance. Her fifteen-month journey around the world became a thing of legend. In fact, it had all the hallmarks of a theatrical production, in which she played the starring role, and the world was her stage. Annie didn't just spin the wheels of her bicycle on her globetrotting mission, she also managed to spin an entire adventure around her sensational feat. She was 'an inveterate story-teller, consummate self-promoter, and masterful creator of her own myth,' turning her journey 'into one of the most outrageous chapters in cycling history, and herself into one of the most colorful characters of the gay 1890s.'[1]

Annie's one-woman crusade for fame and fortune earns her a place in a niche category amongst her exploring counterparts. Whilst her contemporaries were quietly going about the business of exploring and travelling for the sheer joy of the experience, Annie was busy selling autographed photographs of herself and tapping up businesses for her next sponsorship deal. In some respects, it was a stroke of genius. Putting aside the fact that she would have to pedal thousands of miles to achieve her goal – fame and fortune – for a confident and intelligent young opportunist, it was the perfect idea. Whether it was Annie's idea or not is up for debate. She professed that she was making the trip to settle a wager between two wealthy Boston sugar merchants, but the origins of why and how her cycling marathon came about was the first in a long line of slightly dubious claims Annie made over the course of her life. These included the size of the 'prize' money (anywhere from $1000 to $30000 depending on who she was speaking to); the fact the wager prohibited her from getting married during the trip (she was already married with three children when she left Boston); that she had studied medicine for two years and specialized in the 'cultivation of physical beauty' and made money 'dissecting cadavers' (she was a newspaper advertising solicitor); that she was a wealthy heiress who had inherited a 'substantial fortune' (not as far as records show); that she was made 'an orphan at a very young age' (untrue); and perhaps the most grandiose of all, that she was the cousin of a United States congressman and the niece of a United States senator. Annie's biographer and great-grandnephew, Peter Zheutlin gives some insight into Annie's outlandish claims, suggesting that:

> she often seemed to take delight in pulling the legs of reporters, almost all of whom were men at the time, and in testing the limits of their credulity. But the sheer randomness and grandiosity of some of her claims hints at an almost pathological aversion to telling a straight story, though she was never delusional – she knew exactly what she was doing and appears to have enjoyed the game, almost daring reporters to find out who she was.[2]

But despite her inability to tell the truth, or perhaps because of it, one can't help but admire her audaciousness and single-minded commitment to her goal. 'A Yankee for enterprise and daring,'[3] the British press hailed her. Even her name wasn't real. The surname 'Londonderry' was borrowed from New Hampshire's Londonderry Lithia Spring Water company, who asked her to adopt the name as part of her $100 sponsorship deal. Annie's real name was Annie Kopchovsky so her name change served a dual purpose; not only would it promote the business who were sponsoring her, but it also concealed her Jewish origins, easing her journey across different regions and countries.

Annie was also liberal with her interpretation of the 'rules' of the wager and was just as likely to hop onto a train for a stretch, or take a steamship around the coast, as she was to pedal her way around the globe. She had made it to Marseille by 13 January 1895, and was greeted to a hero's welcome and loud cheering from the assembled crowds – particularly as she was (allegedly) pedalling with one foot as the other was bandaged up from an injury and propped on the handlebars. After being (allegedly) held up by French bandits who were most unhappy that she only had three francs in her possession, she departed Marseille on 20 January, and over the next seven weeks rode her Sterling bicycle across northern Africa and the Holy Land, India, China the Korean peninsula and Siberia. She hunted Bengal tigers with German royalty in the India jungle; narrowly avoided being killed by the 'Asiastics' who thought she was an evil spirit; travelled to the front line of the Sino-Japanese War with two journalists and a missionary, where she fell through a frozen river and was shot in the shoulder causing a four-week delay to her journey; rode over fields littered with the war-dead before being thrown into a Japanese prison where she witnessed a Japanese soldier execute a Chinese prisoner. She also managed to squeeze in a trip to Siberia, where she also found time to study the Russian penal system, particularly how it dealt with its political prisoners. She then departed from Yokohama for the final sea-bound leg of her journey on 9 March 1895.

Considering the astonishing distance that she needed to cover on her bicycle, and the alleged month-long delay due to her bullet wound, Annie did exceedingly well to cram in so many miles and so many escapades on the seven week stretch between France and Japan.

A little too well, in fact. 'While all of this high adventure in the jungles of India, at the war front in China, and in the bitter cold of Siberia was unfolding in her fertile imagination,' writes Zheutlin, 'Annie was instead onboard the *Sydney*, skipping like a stone across the water.'[4]

Inevitably, Annie's tall tales began to catch up with her. 'A lady who aspires to the feat of girdling the globe on a bicycle must be the possessor of an unlimited supply of nerve and assurance,' remarked the *Penny Illustrated Paper*, '…up to the present, however, the bicycle hasn't given her much trouble – sea trips seem more in her line.'[5] The *Singapore Straits* were even more cynical: 'The cyclist was introduced as Miss Londonderry and when she spoke one would have known at a considerable distance that she hailed from the land of the stars and stripes…And when she took reduced passage from Marseilles to Yokohama there must have been 50,000 people on hand to give her a royal farewell. But we fancy she exaggerates. Some women do.'[6]

And their caustic assessment of 'Miss Londonderry' didn't stop there. 'Singapore has been reached, therefore, by the easy method of a French mail [the steam ship, *Sydney*] on the cheap...To the chagrin of the wagerers she will be in Boston once more...with the $5000 legitimately earned according to the terms of the contract.'[7]

Annie may genuinely have travelled to some of the destinations she mentioned to the press. She wrote in at least one American newspaper that she had visited Jerusalem, and her lecture tour featured lantern slides depicting the Holy City – although Annie herself was notably absent from them. The steam ship itself made several stops, to take on board coal and other provisions and Annie would have disembarked and toured the port towns and cities to generate interest in her adventure and sell souvenirs. She was a one-woman cycling PR machine, and without making a concerted effort to keep herself in the spotlight, all her endeavours would be for nothing. But that was the thing with Annie – the cycling aspect of the adventure was really a means to an end – the real motivating factor was the fame and fortune awaiting her on her return to America. So, if she needed to embellish some of her antics to gain column inches, then that was fair game to Annie. As was adding a little extra colour to her lecture tour – because spending seven weeks sitting aboard a boat or watching the world whizz by out of a train window just wasn't that interesting to listen to or read about. And it certainly wasn't going to win her any fans. She was skilled at blurring the lines between fiction and reality, and in an age before package tourism, mass media and internet fact checking she could rely on a certain amount of public naivety to obscure the truth.

Nine months after she left, Annie arrived back in America – California to be precise. 'Miss Annie Londonderry has covered 28,000 miles in her unique ride, and has met with several thrilling experiences,'[8] announced *American Cycling*. According to the same newspaper, these included being robbed by a highwayman, being run over by another cyclist in California which resulted in a five week stay in hospital, and a bad fall which resulted in a broken arm (which she then cycled 175 miles with). 'For pluck and endurance such a ride cannot beaten this side of the herring pond,' boasted the press.

But for all the hype, there was no fanfare or welcoming committee when she did finally step onto American soil. Annie was disappointed, and she was quick to say so, telling the local press that she was 'somewhat disappointed that the populace, headed by several brass bands, had not turned out to welcome her as had been the case, according to her diary, in other great cities.'[9] She would have to work harder to win over her fellow country-men and ladies. Her cause was not helped by the 'cutting edge' technology she travelled with – the cyclometer (an odometer for bicycles) attached to her Sterling bike. Annie's 'unique ride' was nowhere close to the 28,000 miles being reported in the press. She had travelled some distance close to that, but her miles cycling i.e., when she was actually spinning the wheel under her own steam, came in at a rather pathetic 7,280 miles, eventually rising to 9,604 miles once she had ridden the final leg of her journey from San Francisco to Boston – which is entirely consistent with the distances involved (barring a few train rides) up to the point she got on the steamer at Marseille. Scenting blood, the press changed their tune. 'Smart girl, Annie Londonderry,' wrote *Cycling Life*, 'but much too fresh to be touring the world as a representative of American womanhood...To read of the hair-breadth escapes of this young woman, to say nothing of the hair-breadth escapes of the people who make her acquaintance, makes our eyes bulge with astonishment.'[10]

The cycling community also joined the throng of criticism. Tom Winder, an American newspaper editor undertaking a similar cycling feat as Annie wrote that, 'Miss Anna [sic] is a hustler for sure...She intends to write a book, so I thus early put the public on its guard.'[11] And the criticism wasn't confined to her homeland either: 'She is not elevating either cycling or her sex,' wrote one rather scathing British cycling magazine.

But in some respects, they were wrong. Annie Londonderry had pulled off an even greater PR stunt; one that went far beyond fame or fortune; because Annie had been smart enough to recognise what cycling represented to women at the end of the nineteenth century, and she was able to - quite literally - ride on the zeitgeist of women's lib. During her nine-month absence from America, she was able to reframe herself as a 'New Woman' and become an outspoken and potent icon of female equality. During the final leg of her journey in 1895 she spoke expansively about the physical and mental benefits of cycling for women, as well as promoting riding without the constraints of corsets or 'heavy, baggy bloomers that make the work a torture.' Instead, she advised: 'A heavy sweater, a neat pair of bloomers, leggings and a natty cap constitute proper costume.' Annie was flying in the face of convention with her wise words. Many women had already been discouraged from cycling for fear of cultivating a 'bicycle face' – a false 'medical' condition dreamed up by a man called Dr Shadwell, to stop women from getting on a bike and pedalling themselves away to freedom. The 'condition' manifested itself thus: 'a peculiar strained, set look so often associated with this pastime... eyes fixed before them, and an expression either anxious, irritable, or at best stony, they pedal away.'[12] In what reads as a thinly veiled argument against women's lib, the doctor smugly asserts that, 'Some wear the 'face' more and some less marked, but nearly all have it... Has anybody ever seen persons on bicycles talking and laughing and looking jolly, like persons engaged in any other amusement? Never, I swear.'[13] Dr Shadwell had obviously never met Annie Londonderry. Fortunately, the eminently sensible Chicago doctor Sarah Hackett Stevenson finally put the matter to rest in 1897 by stating to the *Phrenological Journal and Science of Health* that, '[Cycling] is not injurious to any part of the anatomy, as it improves the general health. I have been conscientiously recommending bicycling for the last five years.'[14]

Annie was spot on with her fashion advice too. The bicycle craze of the late nineteenth and early twentieth century helped to stoke dress reform movements and free women from their corsets so they could bend, stretch, run, swim, cycle, and do all the other things men had been enjoying for centuries. Little wonder the humble bicycle was seen as an instrument of feminism. It was the ultimate symbol of mobility; providing a means for women the world over to escape domesticity and go out on their own unchaperoned. This new found autonomy, courtesy of the bike, was the perfect partner for the women's suffrage movement and cycling was eagerly taken up by its supporters. 'To men, the bicycle in the beginning was merely a new toy,' declared *Munsey's Magazine* in 1896, 'another machine added to the long list of devices they knew in their work and play. To women, it was a steed upon which they rode into a new world.' Even Lillias Campbell Davidson was enthusiastic about cycling. Six years before Annie embarked on her travels, Campbell Davidson dedicated an entire chapter of *Hints to Lady Travellers* to cycling tours and was keen to affirm that, 'No style of touring is at once so enjoyable and so health-giving,' and that new riders could look forward to an activity that would 'raise her spirits and exhilarate her pulses' - although it came with a proviso that, 'Cycling for women has only two real perils – that of over-fatigue or of catching a chill when overheated...'[15] Campbell Davidson went on to establish the Lady Cyclists' Association in 1892 and wrote the *Handbook for Lady Cyclists* which was published in 1896 and promoted, 'A new world of enjoyment,' which would be 'unlocked to the woman who finds herself a-wheel...cycling is a door that leads to many paths of pleasure.'[16]

For Annie, that pleasure was fame and notoriety. But there was also a lot of hand wringing going on behind the scenes. 'Bicycling by young women has helped to swell the ranks of reckless girls who finally drift into the standing army of outcast women in the United States,' wrote Charlotte Smith of the Women's Rescue League, in 1896. To impose a modicum of safety and control over all these free-wheeling feminists, cycling 'rules' were frequently printed in newspapers, particularly in America. In the same year that Annie completed her journey around the globe, the *New York World* published '41 rules for female cyclists' which included such nuggets of advice as, 'Don't refuse assistance up a hill,' 'Don't criticise people's legs,'

'Don't emulate your brother's attitude if he rides parallel toward the ground,' and 'Don't discuss bloomers with every man you know.'

For shame!

But as for Annie… as history had already shown, she was never one for conforming. After her trip, which she completed on 12 September 1895, she turned her attention to journalism, writing with all the aplomb that had made her a celebrity, and with the by-line, 'The New Woman.' 'I am a journalist and a 'new woman,'' she wrote in her first article, 'if that term means that I believe I can do anything that any man can do.'[17]

* * * * *

Six years before Annie Londonderry cycled her way around the world (allegedly), in 1889 another plucky young American lady was making similar plans. Elizabeth Cochran, better known as Nellie Bly, was about to embark on her own record-breaking around-the-world attempt. This time, the feat *was* taken straight from the pages of a fictional story – Jules Verne's *Around the World in Eighty Days* (1873). The journey would require travelling almost 25,000 miles. At 9.40am on 14 November 1889, Nellie boarded the steamer *Augustus Victoria* to start the first leg of her journey across the Atlantic from New Jersey to Southampton, England.

Nellie had been a newspaper journalist for *The World* in New York for two years and had already garnered acclaim for her investigative reporting and the sensational exposé *Ten Days in a Mad House* (1887) – a series of articles that cast an unpleasant light on the treatment of female inmates at the Women's Lunatic Asylum on Blackwell's Island, New York. Bly was one of the first in a growing number of 'stunt' or 'undercover' female reporters – thought to be the brainchild of publisher and owner of *The World,* Joseph Pulitzer. However, Nellie stood out from other journalists from the beginning through her audacity and at times fearless approach to her reporting. Her numerous adventures as a roving reporter ushered in a new era of journalism – one which highlighted social injustice and the plight of the common (and often poor) man, as well as a hefty dose of less serious reporting that would capture the public's imagination and make a light-hearted but entertaining read. She trained with a boxing champion; went underwater in a diving bell; went up in a balloon; performed as a chorus girl at the Academy of Music – with limited success because she forgot her cue to exit and ended up standing alone onstage; and she also met a young Helen Keller, the future disability rights advocate and political activist. She even bought a baby to prove the ease at which the white slave trade operated in New York. In an article titled 'Nellie Bly Buys a Baby' she exclaims indignantly: 'I bought a baby last week, to learn how baby slaves are bought and sold in the city of New York. Think of it! An immortal soul bartered for $10…Slavery exists today in New York in a more repulsive form than it ever existed in the South. White slaves, baby slaves – young, innocent, helpless baby slaves – bought and sold every day in the week – bargained for before they are born – sold by their parents!'[18]

Her job inevitably brought with it a set of risks. Never one to shy away from hard-hitting reportage and divisive subjects, her first journalist posting at the *Pittsburgh Dispatch* saw her writing about how divorce affected women and presented an argument for divorce law reform. From this, she moved on to the plight of women factory workers, but this angered local factory owners, so Nellie was relegated to the women's pages to cover the softer and more 'gender appropriate' subjects – fashion, society and gardening. In frustration at what she must have seen as discrimination she decided to try something completely different. 'I was too impatient to work along at the usual duties assigned women on newspapers,' she wrote, 'so I conceived the idea of going away as a correspondent.'[19] Her chosen location was Mexico, then a dictatorship under Porfirio Diaz, which she planned to explore as a foreign correspondent; reporting on the lives and customs of the Mexican people.

'It is not a clean, inviting crowd, with blue eyes and sunny hair I would take you among,' she wrote in *Six Months in Mexico*, 'but a short, heavy-set people, with almost black skins, topped off with the blackest eyes and masses of raven hair. Their lives are as dark as their skins and hair and are invaded by no hope that through effort their lives may amount to something.' It was a damning assessment, but she directed her criticism for the plight of the Mexican people squarely at their government and its tyrannical and undemocratically elected leader, Diaz:

> The constitution of Mexico is said to excel, in the way of freedom and liberty to its subjects, that of the United States; but it is only on paper. It is a republic only in name, being in reality the worst monarchy in existence. Its subjects know nothing of the delights of a presidential campaign; they are men of a voting age, but they have never indulged in this manly pursuit, which even our women are hankering after. No two candidates are nominated for the position, but the organized ring allows one of its members – whoever has the most power – to say who shall be president; they can vote, though they are not known to do so; they think it saves trouble, time, and expense to say at first, 'this is the president,' and not go to the trouble of having a whole nation come forward and cast the votes…this is the ballot in Mexico.[20]

But particularly galling for Nellie was the complete lack of a free press, and that most of its citizens – some eight million, Nellie suggests – were completely disenfranchised. 'This large majority has no voice in any matter whatever,' she wrote, 'so the government is conducted by the smaller, but so-called better class…the Mexican papers never publish one word against the government or officials, and the people who are at their mercy dare not breathe one word against them, as those in position are more able than the most tyrannical czar to make their life miserable.'[21] She was horrified to discover that the Mexican press was in the pocket of the government and 'paid to abstain from attacks on the government.'

It didn't end there, however. Editors of the Mexican newspapers weren't just despised by the Mexican people, they were risking their lives if they spoke out about their government. 'If a newspaper even hints that government affairs could be bettered, the editors are thrown into prison, too filthy for brutes, until they die or swear never to repeat the offense. The papers containing the so-called libelous items are all hunted up by the police and destroyed, and the office and type are destroyed. These arrests are not unusual; indeed, they are of frequent occurrence.'[22] This persecution, Bly recounts, caused one unfortunate editor to die from his treatment in jail, and another to be imprisoned for reporting on a judicial ruling involving the secret police. Citizens were also scrutinized lest they say anything libelous – which if they did, meant the same tragic outcome as members of the press. 'One fellow who ran a liquor shop let his tongue wag too much for wisdom,' Bly narrates, 'and one night a member of the police secret service went in, and as the proprietor turned to get the drink the policeman had called for, he was shot in the back and again in the body after he had fallen.'[23]

As well as what she learned about the press, Nellie also wrote about the daily lives of the people of Mexico, describing courtship, wedding customs, drinking, the popularity of tobacco and their addiction to playing the lottery, noting that 'Crippled, blind, aged, poverty-stricken men and women are on the streets at all hours selling numbered strips of tissue paper marked 'Lottery'.' At one point she was approached by a grey-haired lottery woman, who said to her in Spanish: "Buy a ticket. A sure chance to get $10,000 for twenty-five cents.' Being told that we had no faith in lotteries, she replied: 'Buy one; the Blessed Virgin will bring you the money."[24]

But the morning of 14 November 1889 – the day she began her around the world trip – was her biggest and most sensational adventure of all. She had already raised the stakes, claiming that unlike the title of Jules Verne's novel, on which her journey was based, she would be aiming to complete the trip in just seventy-five days. She had just three days to prepare herself because despite suggesting the idea a year earlier, her editor at *The World* didn't want her to travel unchaperoned (at the 'tender age' of twenty-four). She related the conversation with her editor in her travelogue, *Around the World in Seventy-Two Days* (1890):

> 'It is impossible for you to do it,' was the terrible verdict.
> 'In the first place you are a woman and would need a protector, and even if it were possible for you to travel alone you would need to carry so much baggage that it would detain you in making rapid changes. Besides you speak nothing but English, so there is no use talking about it; no one but a man can do this.'[25]

Spirited Nellie wouldn't give up without a fight however, so she resorted to a little blackmail to get her desired outcome.

'Very well,' I said angrily. 'Start the man, and I'll start the same day for some other newspaper and beat him.'[26]

They clearly underestimated her, but no doubt this only fuelled Nellie's commitment to her goal even more. It must have done the trick because she writes next that, 'I would not say that this had any influence on their decision, but I do know that before we parted I was made happy by the promise that if any one was commissioned to make the trip, I should be that one.'[27]

The British press didn't quite know what to make of Nellie Bly. 'The extent to which women are employed on American papers is incredible in England,' wrote the *Luton Reporter* in 1889, 'where the newspapers are filled more with fact than with fancy, and yet they can't find material enough in their own country to supply the 'woman's page' in their huge Sunday papers.'[28] Branded an 'Eccentric Young Man-Hater,'[29] by *The Leeds Times,* of Nellie herself they were similarly uncomplimentary: 'Nellie Bly writes a weekly article for the *World* of very bad literary quality, and more of less sensational in tone, yet she is paid at the rate of £500 a year, with expenses.'[30] No sour grapes there then…

British sensibilities weren't quite ready for such sensationalist reporting or for such intrepid go-getters as the likes of Nellie - for Nellie wasn't the only journalist embarking on a world tour that day in 1889. She also had competition from another female journalist - Elizabeth Bisland of the *Cosmopolitan Magazine*. Described by *Irish Society* as 'quite a different sort of woman' to Nellie; 'refined, rather retiring and an altogether reflective little body'[31]; Elizabeth planned to take the opposite route to Nellie, starting out westwards via San Francisco, Japan, China and India and the Suez Canal. In comparison to Nellie's three days of preparation, Elizabeth Bisland's editor gave her just fifteen minutes – asking her the same afternoon that Nellie departed, to which Elizabeth 'replied that she could, and caught that evening's New York Central train for the West.'[32]

The race was on.

Except it wasn't because Nellie wasn't prepared to stoop to such copycat tactics. 'I would not race,' she said. 'If someone else wants to do the trip in less time, that is their concern. If they take it upon themselves to race against me, it is their lookout that they succeed. I am not racing. I promised to do the trip in seventy-five days, and I will do it...'[33]

Nellie only learnt of her rival thirty-nine days after leaving New York, once she had arrived in Hong Kong, a city she described as 'strangely picturesque.'[34] In spite of her composure, which *The World* was keen to publicise – pronouncing the day after she left America that she had shown 'not a wince of fear or trepidation'[35] on her departure - Nellie does nothing to disguise her nervousness in her own account of the trip, relating with honesty the anxiety she was feeling about travelling on her own and her experience of 'strange lands and strange people.' 'I felt lost,' she wrote about leaving America for the first time. 'My head felt dizzy and my heart felt as if it would burst. Only seventy-five days! Yes, but it seemed an age and the world lost its roundness and seemed a long distance with no end, and–well, I never turn[ed] back. 'I am off,' I thought sadly, 'and shall I ever get back?'[36] Fear of the unknown was Nellie's biggest challenge, which was exacerbated by her well-meaning friends and family, who had 'helpfully' given her an idea of what she could expect: 'Intense heat, bitter cold, terrible storms, shipwrecks, fevers, all such agreeable topics had been drummed into me until I felt much as I imagine one would feel if shut in a cave of midnight darkness and told that all sorts of horrors were waiting to gobble one up.'[37]

Unfortunately for Nellie, her initial crossing from New York to Southampton on the *Augusta Victoria* was a baptism of fire. She had never been on a sea-voyage before and had set out at the worst time of year when the ferocity of storms over the Atlantic were at their peak. So, it wasn't long before the rollicking waves caught up with her. 'I could expect nothing else than a lively tussle with the disease of the wave,' she wrote confidently.

> 'Do you get sea-sick?' I was asked in an interested, friendly way. That was enough; I flew to the railing. Sick? I looked blindly down, caring little what the wild waves were saying, and gave vent to my feelings. People are always unfeeling about sea-sickness. When I wiped the tears from my eyes and turned around, I saw smiles on the face of every passenger. I have noticed that they are always on the same side of the ship when one is taken suddenly, overcome, as it were, with one's own emotions.
>
> The smiles did not bother me, but one man said sneeringly: 'And she's going around the world!'
>
> I too joined in the laugh that followed. Silently I marveled at my boldness to attempt such a feat wholly unused, as I was, to sea-voyages. Still I did not entertain one doubt as to the result.[38]

The South China Sea had more adventures in store for Nellie later in her trip. On route from Singapore to Hong Kong the seasickness caught up with her again, although this time she at least took some pleasure in the cause of her discomfort, writing that: 'The terrible swell of the sea during the Monsoon was the most beautiful thing I ever saw. I would sit breathless on deck watching the bow of the ship standing upright on a wave then dash headlong down as if intending to carry us to the bottom.'[39] Nellie soon changed her mind though when the monsoon threatened to engulf them all.

> One night during the monsoon the sea washed over the ship in a frightful manner. I found my cabin filled with water, which, however, did not touch my berth. Escape to the lower deck was impossible, as I could not tell the deck from the angry, pitching sea...I thought it very possible that I had spoken my last word to any mortal, that the ship would doubtless sink, and with it all I thought, if the ship did go down, no one would be able to tell whether I could have gone around the world in seventy-five days or not. The thought was very comforting at that time, for I felt then I might not get around in one hundred days.[40]

Besides her adventures on the high seas, it is Nellie's depiction of the people she met – other travellers and native countrymen and women - that makes her account so authentic and engaging. At times gossipy – 'Talk about cranks! One woman told the chief officer one day that she wanted a cabin just over the ship's screw so she could tell that the ship was going!'[41] - and always revealing - 'I never knew how queer our passengers were until we reached Hong Kong...When we landed, a man sued the company for getting him in ahead of time. He said he bought his tickets to cover a certain length of time, and if the company got him in before it expired they were responsible for his expenses, and they had to pay his hotel bill,'[42] - Nellie's grasp of the human character, particularly her study of other passengers, is just as authentic as one of her ethnographic exploring sisters. Although of course, she doesn't stop to think how 'queer' her own travelling plans must have appeared to other people. That is all part of her charm though.

One person she certainly wouldn't have appeared 'queer' to was the writer whose novel inspired her trip. Nellie was fortunate enough during the European leg of her journey to meet Jules Verne, describing his snow-white hair as 'rather long and heavy' and 'standing up in artistic disorder.' 'The brilliancy of his bright eyes,' she wrote, 'and the rapidity of his speech and the quick movements of his firm white hands all bespoke energy–life–with enthusiasm.'[43] During their meeting she was surprised to learn the origins of the idea for his most famous work of fiction. Ever the journalist, Nellie couldn't help but interview Verne, asking him outright, "How did you get the idea for your novel, *Around the World in Eighty Days*?' Verne replied that, 'I got it from a newspaper,'

> I took up a copy of 'Le Siècle' one morning, and found in it a discussion and some calculations showing that the journey around the world might be done in eighty days. The idea pleased me, and while thinking it over it struck me that in their calculations, they had not called into account the difference in the meridians and I thought what a denouement such a thing would make in a novel, so I went to work to write one. Had it not been for the denouement I don't think that I should ever have written the book.[44]

That the celebrated book on which her journey was based had originated from some humble journalism in a newspaper no doubt validated Nellie's belief in her trip even further. That she, Nellie Bly, the journalist born in Pittsburgh, Pennsylvania was attempting to make real a fictionalized account by an acclaimed French author was all the encouragement she needed – and unlike Verne's protagonist Phileas Fogg, she was also a woman – and there was a certain amount of cachet attached to that too. She would later tell the chief engineer on one of the ships she sailed on that she would rather die than arrive late in New York – and she meant it too. 'If I fail, I will never return to New York,' she wrote. 'I would rather go in dead and successful than alive and behind time.'[45]

Nellie Bly was many things, but she certainly wasn't a loser.

Her accounts of her adventures on land are filled with all the curiosity and wonder of someone confronted with something new and alien. In Aden (now in Yemen), Nellie was entranced by the women who 'walked proudly along, their brown, bare feet stepping lightly on the smooth road. They had long purple-black hair, which was always adorned with a long, stiff feather, dyed of brilliant red, green, purple, and like striking shades. They wore no other ornament than the colored feather, which lent them an air of pride, when seen beside the much-bejeweled people of that quaint town. Many of the women, who seemed very poor indeed, were lavishly dressed in jewelry.'[46] In awe of their beauty, she writes, 'To me the sight of these perfect, bronze-like women, with a graceful drapery of thin silk wound about the waist, falling to the knees, and a corner taken up the back and brought across the bust, was most bewitching.'[47] She was just as enamoured with her surroundings. In Ceylon (now Sri Lanka) she was 'impressed with the beauties of Colombo.'

> As we moved in among the beautiful ships laying at anchor, we could see the green island dotted with low arcaded buildings which looked, in the glare of the sun, like marble palaces. In the rear of us was the blue, blue sea, jumping up into little hills that formed into snow drifts which softly sank into the blue again…beyond a wide road ran along the water's edge until it was lost at the base of a high green eminence that stood well out over the sea, crowned with a castle-like building glistening in the sunlight.[48]

Nellie's only failing is that she only has time to describe what she sees (which of course, as a journalist, she does well) instead of attempting to understand the culture and customs of the foreign lands she travels through. As one English newspaper commented, 'There would be no time to take any notice of the places visited, and the report of the adventure would, we should imagine, be rather dry reading.'[49] In reality, Nellie's travelogue is anything but that – it just has entertainment at its heart rather than any high-minded designs to further the fields of science or ethnography. When she does describe the native rituals she observes, it is with the fascination of a tourist. In Tokyo, she 'saw the most disreputable looking god. It had no nose. The Japanese believe if they have a pain or ache and they rub their hands over the face of that god, and then where the pain is located, they will straightway be cured. I can't say whether it cured them or not, but I know they rubbed away the nose of the god.' Such a description is typical of Nellie's light-hearted but authentic narrative. She told things how they were, in a way that she knew her readership would understand – which is why Nellie Bly's exploits sold papers.

She certainly benefited from a more welcoming homecoming than Annie Londonderry had experienced five years later. When she finally docked at San Francisco and disembarked from the *Oceanic,* she was overwhelmed by the reception given to her. To her fellow Americans she was a heroine, and they 'turned out to do honor to an American girl who had been the first to make a record of a flying trip around the world, and I rejoiced with them that it was an American girl who had done it.'[50] She remembered her whistle-stop trip across the continent, from San Francisco to New York, as:

> one maze of happy greetings, happy wishes, congratulating telegrams, fruit, flowers, loud cheers, wild hurrahs, rapid hand-shaking and a beautiful car filled with fragrant flowers attached to a swift engine that was tearing like mad through flower-dotted valley and over snow-tipped mountain, on–on–on! It was glorious! A ride worthy a queen. They say no man or woman in America ever received ovations like those given me during my flying trip across the continent.[51]

Explorer Isabella Bird Bishop (1831-1904) in 1899 wearing a Manchurian gown. Isabella is perhaps one of the more well-known lady explorers of the nineteenth-century and was the first female fellow of the Royal Geographical Society.

American journalist Elizabeth 'Nellie' Bly (1864-1922), in a publicity photo for her around-the-world voyage which she began in November 1889. The trip took her 72 days, 6 hours, and 11 minutes.

Wealthy heiress and socialite Aimée Crocker (1864 -1941) in 1890. Aimée scandalised late nineteenth century society with her exploits abroad and 'devil may care' attitude to life.

Mary Brodrick (1858-1933), photograph date unknown. Mary was one of the first female scholars to study Egyptology.

Left: Ida Laura Pfeiffer (1797-1858) in 1856. Of all the lady explorers of the era, Ida travelled the furthest – an estimated 32,000 km during the final sixteen years of her life.

Below: May French Sheldon's (1847-1936) palanquin, which she renamed her 'white elephant.' She travelled across Africa with 150 Zanzibari porters and guides. Image courtesy of the Wellcome Trust.

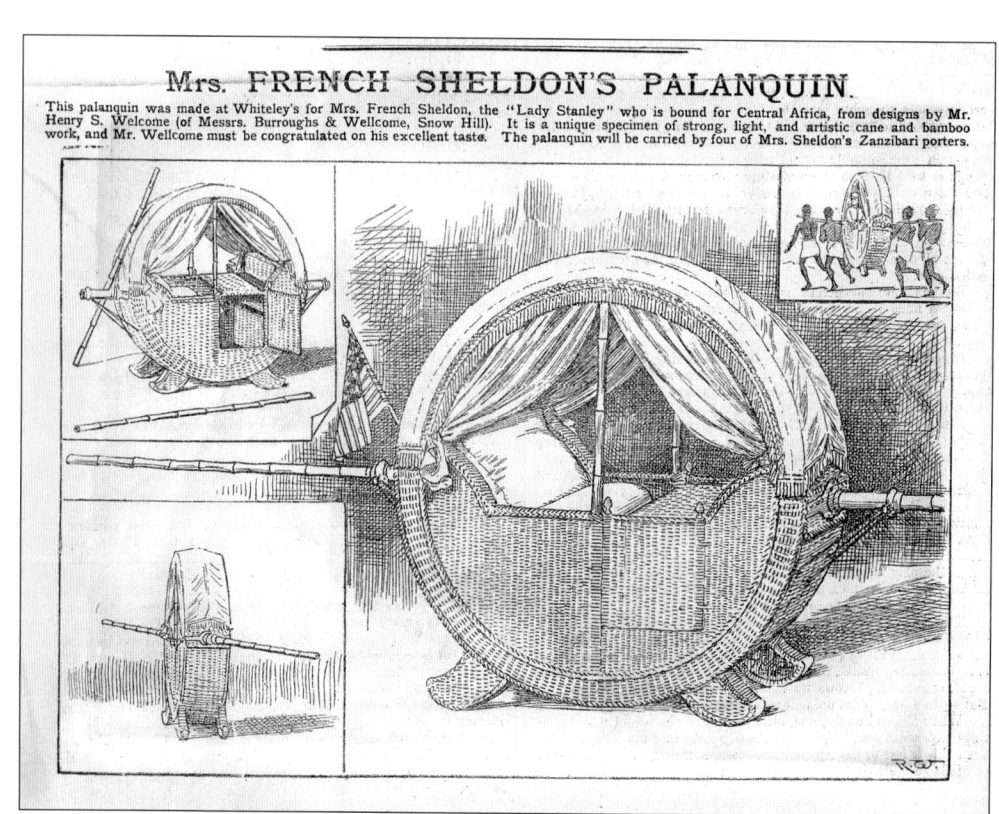

Mrs. FRENCH SHELDON'S PALANQUIN.

This palanquin was made at Whiteley's for Mrs. French Sheldon, the "Lady Stanley" who is bound for Central Africa, from designs by **Mr. Henry S. Welcome** (of Messrs. Burroughs & Wellcome, Snow Hill). It is a unique specimen of strong, light, and artistic cane and bamboo work, and Mr. Wellcome must be congratulated on his excellent taste. The palanquin will be carried by four of Mrs. Sheldon's Zanzibari porters.

American explorer and ethnographer May French Sheldon (1847-1936). Image courtesy of the Wellcome Trust.

Catharine Parr Traill (1802-1899) in 1884, author and botanist who wrote one of the earliest descriptions of pioneer life in Upper Canada. Image courtesy of Library and Archives Canada, PA-802715.

Above: Canadian missionary Dr. Susie Rijnhart (1868-1908) in Tibetan dress in 1901. Susie was the second Western woman known to have visited Tibet, after Annie Royle Taylor. She faced extremely challenging circumstances during her first journey but never lost her faith.

Left: Lady Jane Franklin née Griffin (1791-1875) in 1816 by Amelie Romilly (1788-1875). Jane was a seasoned traveller and colonial wife but her greatest triumph was championing the memory of her late husband's Arctic explorations.

Margaret Fountaine (1862-1940) in 1886. Margaret's biggest passions were butterflies and men – both of which she found on her travels in the latter half of the nineteenth-century.

Marianne North (1830-1890) in a familiar pose, taken in 1886. Marianne challenged the accepted method for painting plants and botanical subjects and travelled the world to represent as many species as she could.

Annie Cohen Kopchovsky aka Annie Londonderry (1870-1847) in 1896. Annie gained notoriety for 'circumcycling' the globe, despite using other modes of transport such as trains and steam ships to help her along the way.

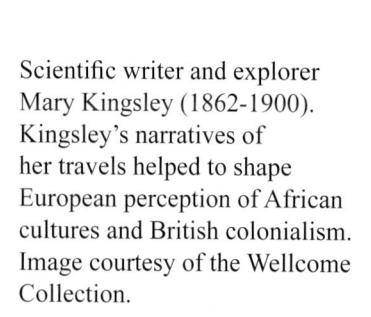

Scientific writer and explorer Mary Kingsley (1862-1900). Kingsley's narratives of her travels helped to shape European perception of African cultures and British colonialism. Image courtesy of the Wellcome Collection.

Lady Hester Stanhope (1776-1839) on horseback – a sketch taken from her memoirs, which were written after her death by her physician Dr Meryon, who had accompanied her on her travels. Image courtesy of the Wellcome Collection.

Lady Hester Stanhope in Eastern clothes. Hester settled in the East and made her home there. She never returned to England, although her adventures were often discussed in the British press.

Left: British missionary and explorer Kate Marsden (1859-1931) in full travelling dress with a map of her route through Siberia, from her travelogue *On Sledge and Horseback to Outcast Siberian Lepers* (1892). Kate's outfit kept her warm but made it impossible for her to bend; 'a decided burden in more ways than one,' she later wrote.

Below: Kate Marsden travelling in Siberia. Along with Isabella Bird Bishop, Kate was one of the first women to be elected as a fellow of the Royal Geographical Society.

Annie Royle Taylor (1855-1922), English explorer and Evangelical missionary. The determination she showed in defying her father to undertake missionary work abroad would come in useful for the challenges she faced in the field. From the pamphlet *The Tibetan Pioneer Mission* (1894).

MISS ANNIE TAYLOR AND HER TIBETAN SERVANT.

Caroline (Cara) Edgeworth David (1856-1951), from *Funafuti, or, Three months on a coral island: an unscientific account of a scientific expedition* (1899). Cara accompanied her more famous husband, geologist Sir Edgeworth David on a coral hunting expedition to Funafuti (now Tuvalu) and wrote a lively account of their day-to-day life on the island. Image courtesy of the State Library of New South Wales, DSM/988.9/1.

Above left: Lady Florence Baker (1841-1916), a keen traveller who acted as scientific assistant and record keeper to her husband Sir Samuel White Baker during their explorations of Africa.

Above right: Lady Florence Baker with her husband Samuel. Sam Baker rescued Florence from a slave market but their relationship was very much on equal terms.

A depiction of the 'New Woman' from 1901. The stereotypical 'New Woman' wore knickerbockers, smoked cigarettes and rode a bike. Meanwhile, the man is relegated to doing the laundry.

Swiss explorer Isabelle Eberhardt (1877-1904) in 1900. Born into Genevan aristocracy, Eberhardt embraced a nomadic existence and converted to Islam.

'The Bicycle Suit' – a cartoon from the January 1895 issue of *Punch*. The dialogue reads,

Gertrude: "My dear Jessie, what on earth is that Bicycle Suit for?"

Jessie: "Why, to wear, of course."

Gertrude: "But you haven't got a Bicycle!"

Jessie: "No: but I've got a Sewing Machine!"

The bicycle was the ultimate instrument of feminism, as well as sparking an entirely new trend in ladies' fashions.

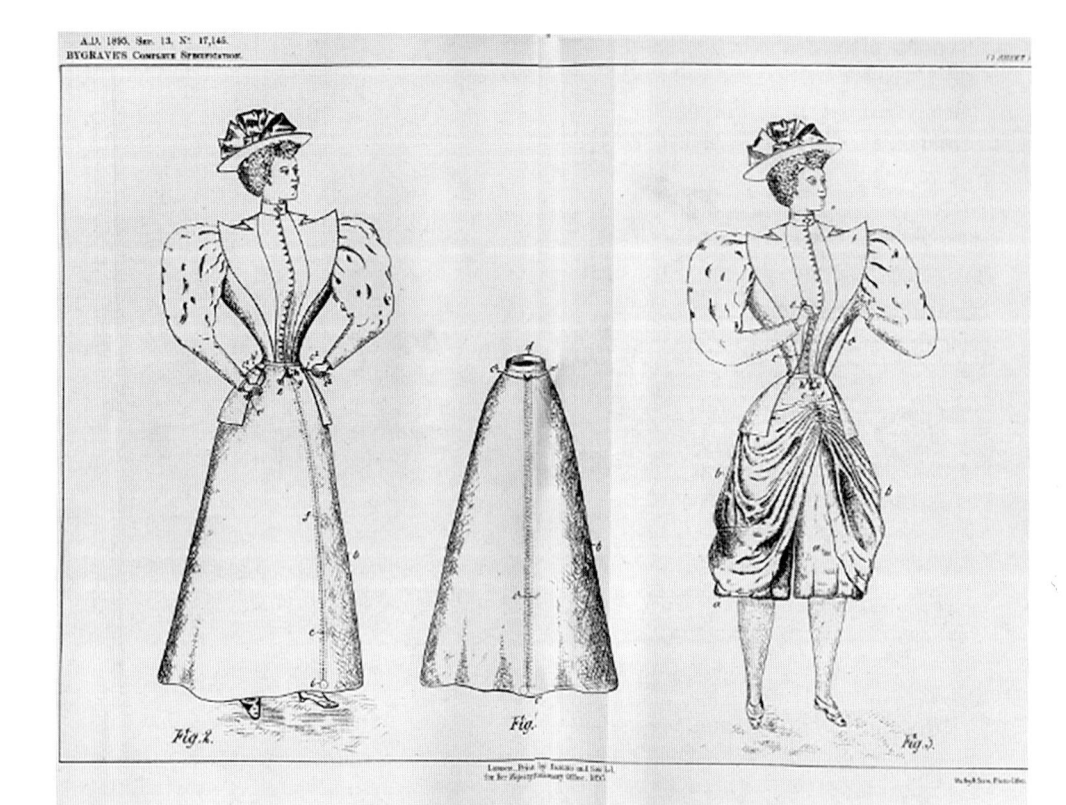

Part of dressmaker Alice Bygrave's patent application for 'Improvements in Ladies' Cycling Skirts,' registered in 1895. Bygrave's invention meant that when the rider was preparing to mount the bicycle, they would pull on two cords that would draw the skirt up at the front and back, revealing the knickerbockers underneath. A year after the patent was filed, the design was commercialised and distributed by the British fashion house Jaeger.

The perilous climb to Chilkoot Pass in February 1899, from *Munsey's Magazine*, made by Martha Black (1866-1957). With the promise of gold aplenty, many prospectors made the terrifying journey to the Klondike gold fields at the turn of the twentieth century.

Right: Prospectors gathered at the bottom of the Chilkoot Pass in 1897/1898. The Canadian government required each person to bring enough food and equipment to last a year. This generally weighed a thousand pounds or more, and it took several trips to carry it over the pass.

Below: Jane Elizabeth Digby, Countess of Ellenborough (1807-1881) – a watercolour by James Holmes (1777-1860). Jane scandalised Victorian society by falling in love with, and marrying, a Bedouin sheik.

Above left: Mary Moffat Livingstone (1820-1862). As the wife of the famous Scottish missionary and African explorer, David Livingstone, her lot was harder than most. They married in 1845 and had six children, some of whom were either born in the wilderness or trekked through it alongside their parents, often at great cost.

Above right: Marianne North in Julia Margaret Cameron's house in Ceylon, photographed by Cameron in 1877. Cameron was a renowned portrait photographer and whilst Marianne thought her portrait 'uninteresting,' she at least had the cachet of knowing she would hang alongside the celebrities of the day such as Charles Darwin and Alfred, Lord Tennyson.

Below: According to the customs of nineteenth century European society, a woman's place was at home where she could attend to her domestic duties, not gallivanting abroad. 'The Angel in the House' – a familiar literary trope taken from Coventry Patmore's poem of the same name – would become the paradigm through which nineteenth century ladies should live their lives.

Luckily for Nellie, *The World* had chartered a train to carry her across the country, no doubt to safeguard their boast that Nellie would arrive home first, but also because it was handy for drumming up even more publicity as 'Nellie Bly's Train.' Along the route, crowds dressed in their Sunday best cheered her on and bands played in her honour. She was asked if she'd ridden an elephant, invited to become a governor of Kansas and shook numerous hands so that her 'arms ached for almost a month afterwards.' One man had his own methods of ensuring Nellie met her eighty-day deadline:

> 'Nellie Bly, you must touch my hand,' he said, excitedly. Anything to please the man. I reached over and touched his hand, and then he shouted:
>
> 'Now you will be successful. I have in my hand the left hind foot of a rabbit!'

But Nellie didn't need tokens of luck. She arrived in New York on 25 January 1890 - a full three days ahead of her own schedule, and eight days before Verne's fictional hero Phileas Fogg. It had taken seventy-two days, six hours, eleven minutes and 14 seconds. Her 'rival' Elizabeth Bisland finally crawled in four and a half days later. She had still beaten Phileas Fogg but she was woefully behind the triumphant 'globe girdler' Nellie Bly. The news of Nellie's return was widely reported – no more so of course than in her own paper, the *New York World* who allegedly dedicated thirty-two columns to her journey – one of which was entirely made up of laudatory headlines such as 'Father Time Outdone!' and 'Thousands Cheer Themselves Hoarse.' 'The *World*, of course, went into hysterics over this not very important event,'[52] wrote one Scottish newspaper, with all the cynicism one would expect from this side of the Atlantic. The *Blackburn Standard* went even further; congratulating Nellie on 'the accomplishment of a wonderful feat,' but also throwing in some casual sexism just to balance the books.

> This performance of Miss Bly will be a great feather in the cap of lady journalists. Here is Miss Nelly Bly round the world and back again in 2½ months. But then this is due no doubt (amongst other reasons) to the way in which women sometimes skim and scamp, compared with the exhaustive deliberation with which the male Commissioner has set himself to study the manners of the men and women whom he has visited.[53]

Nellie's sex inevitably attracted attention – much of this unwanted and unasked for, although Lillias Campbell Davidson would have argued otherwise. Campbell Davidson was adamant that where 'obtrusive attentions from travellers of the other sex,' had been observed, 'the woman has had only herself to blame.'[54] Nellie's account tells a different story. She had started her trips with high hopes, and wrote that 'I knew if my conduct was proper I should always find men ready to protect me, let them be Americans, English, French, German or anything else.'

But her optimism was short-lived. On route from Italy to Egypt she was propositioned by a man who mistakenly thought she was 'an eccentric American heiress, traveling about with a hair brush and a bank book.'[55] Nellie rebuffed him with the same bluntness that he had used to describe his circumstances: 'I found it convenient, later on, to correct the report when a young man came to me to say that I was the kind of a girl he liked, and as he was the second son and his brother would get both the money and the title, his sole ambition was to find a wife who would settle £1,000 a year on him.'[56] On the way to Hong Kong, Nellie found herself in a similar situation. A man who had been 'quite attentive' was struck by seasickness, but was anxious that Nellie got to see him at his best. Nellie writes, "You don't know how nice I can look,' he said pathetically at another time. 'If you would only stay over at Hong Kong for a week you would see how pretty I can look."[57] Even after the man had been (falsely) informed that Nellie was engaged to the chief-officer to stop him following her around 'it only served to increase his devotion.' After an extremely unnerving moment in which Nellie found herself alone on deck with the man and he started to talk wildly about jumping overboard to a watery death with her in his arms, Nellie vowed 'not to spend one moment alone and unprotected on deck.'[58]

Nellie herself wasn't beyond objectifying some of the men she met on her travels. On one occasion, she described a 'good-looking' ship's captain whose 'smooth, youthful face' and 'tall, shapely, slender body'[59] contradicted the 'imaginary short legs, holding upright a wide circumference under an ample waistcoat,'[60] she had been expecting. She also confides some interesting tips for her female readers:

> At every port I touched I found so many bachelors, men of position, means and good appearance, that I naturally began to wonder why women do not flock that way. It was all very well some years ago to say, 'Go West, young man;' but I would say, 'Girls, go East!' There are bachelors enough and to spare! And a most happy time do these bachelors have in the East. They are handsome, jolly and good natured. They have their own fine homes with no one but the servants to look after them. Think of it, and let me whisper, 'Girls, go East!'

Nellie did manage to snag her own man of position and means, although he was thoroughly homegrown in America. Five years after her famous trip, and now attempting the more serious business of novel writing, Nellie married millionaire manufacturing magnate Robert Seaman. He was seventy-three, she was thirty-one. The *Town Tattler* called it a stunt and suggested that Nellie had staged a sham marriage so that she could get an article out of it, but the marriage was real. Based on her own enthusiasm for wealthy bachelors though, it is hard not to be cynical...

Nellie was hot property on her return to America. A board game based on her epic journey was created, and the 'Nellie Bly' style of cap – redolent of the one she had travelled in - quickly became the latest fashion must-have item. Her photograph was sold as a collectible for $5 – a week's wage for a factory girl, and Nellie herself profited from a lecture tour which earned her nearly $10,000. But, unlike Annie Londonderry, Nellie hadn't really done it for the money. She did it because she could - and she approached it with all the gung-ho and spirit with which she approached her other investigative journalism. 'I've always had the feeling,' she wrote, 'that nothing is impossible if one applies a certain amount of energy in the right direction. If you want to do it, you can do it.'

Journalism was her first love though and she would quite literally put her life (and reputation) on the line for her job in the early twentieth century. As early as 1896, Nellie was investigating the Women's Suffrage Movement, covering their Washington D.C. convention for the *New York World* in typical Nellie fashion: 'She had a cold, unsympathetic voice and wore frightful clothes,' she wrote of the Recording Secretary, Alice Stone Blackwell. 'And her black cashmere skirt was shocking. I never saw a skirt hang worse or one more badly made.'[61] However, despite initially being distracted by what everyone was wearing, she gave a thorough account of the proceedings, bringing her customary humour into play ('Charlotte Perkins Stetson has a long name, a large vocabulary, a good voice, an attractive smile and excellent thinking faculties'[62]) Under the column headline, 'Michigan a Great Place for Men,' she records how 'In Michigan married women don't own their own clothes. They belong, with everything else, to the husband. If you live in Michigan and are going to part with your husband, try to get your good clothes out of the house first. If you don't, they'll help him to get a new wife.'[63] Later, when talk at the convention turns to finances, under the subheading 'A Request to the Association', Nellie reports the news that a former supporter, Mrs Doane, left $1000 to the Association in her will:

> This did not rouse any feeling of gratitude that I could perceive, though the burden of their cry had been money.

'Tell all the rich women about to die to go and do likewise, only more so,' commented the President, drily.[64]

Yet hidden amongst the wit, Nellie still manages to convey the crux of the movement. She may have been playful in her reportage, but the Suffrage movement was a topic she returned to on several occasions. Nellie was certainly a feminist and supporter of the movement but never directly put her name to it. She revisited women's suffrage again in March 1913 when she was invited to cover the Woman Suffrage Procession in Washington D.C. for the *New York Evening Journal*. She rode in the procession, alongside over 5000 marchers, floats, bands and groups representing women at home, at school and in the workplace. The article she wrote following the event ran with the inflammatory headline, 'Suffragists Are Men's Superiors,' but she made one prescient comment – that women in the United States wouldn't be given the right to vote until 1920.

The beginning of the Great War in 1914 saw Nellie don a more serious hat. Nellie had initially travelled to Austria to sign her late husband's company over to a wealthy Austrian to avoid it being seized for the large debts it had accrued since his death. But since there was a war on, Nellie reasoned that she may as well put her journalistic talent to use, saying at the time that she 'did not mind at all whether or not she was killed because of her discouragement over her financial affairs.'[65]

Gone was the witty, informal reporting for which she was renowned. As a war correspondent, her writing needed to convey the reality of the conflict she was witnessing so that her fellow Americans could understand the full horror of what was happening in Europe. Nellie was based on the Eastern Front, becoming one of the first female reporters, and certainly the first foreigner, to visit the war zone between Serbia and Austria. Her vivid descriptions of the 'wounded and sick men' were a far cry from the 'frightful clothes' of the suffragists. 'A continuous, straggling stream of sick and wounded soldiers was always coming toward us,' she wrote on 26 December 1914.

> Sometimes they saluted, more often they staggered unconsciously and forlornly on, their sunken eyes fixed pathetically on the west, blind to their surroundings, their ears deaf to the near and ceaseless thundering of cannons, their nerves dead to the awful whizzing of the granards [sic] as they whirled above our heads, so near, yet never visible to the eye.[66]

At fifty, Nellie's perspective had matured. She wanted to make a practical difference to other people's lives, rather than just raise awareness of their plight through her writing. This is no better demonstrated than in her 'Appeal to the American Citizens' via telegram from Austria in May 1915, to 'send packages of wheat and flour by parcel post only for orphans and widows, addressed to Nellie Bly, Vienna, Austria.'[67]

When the war was finally over, she turned her hand to writing an advice column for the *New York Evening Journal*. It was a move away from the high-paced reporting of her younger days which reflected her age and status. She also championed several causes particularly those related to labour and work, and she set up her own adoption agency for babies of unmarried mothers, managing it from her room at the McAlpin Hotel in New York.

She died in the city that had given her fame and the opportunity to further women's journalism. Suffering from pneumonia, Bly quietly took herself off to St Mark's Hospital on 9 January 1922 – the day her last article appeared in the *Journal*. She died on 27 January 1922 and was buried at Woodlawn Cemetery in The Bronx.

'Nellie Bly was the best reporter in America,' her former editor, Arthur Brisbane, wrote the next day, 'and that is saying a great deal. Reporting requires intelligence, precision, honesty of purpose and accuracy…Nellie Bly died too young, cheated of the fortune that should have been her own, suffering for years from ill health that could not diminish her courage or her kindness of heart.'[68]

Nellie was famous in her lifetime, and she remains famous to this day – not because she performed a remarkable round-the-world trip in order to sell newspapers, but because she embodied the spirit of equality and women's rights - raising her fellow lady journalists up from the doldrums of the ladies pages to headlining on the front-cover with columns of engaging and informative copy. Nellie forged a path for herself and for those that followed, but most importantly she did it by staying true to herself. She was authentic, which is patent in her many works and writings – a quality she embodied throughout her life. 'I have never written a word that did not come from my heart,' she professed in her final article for the *Evening Journal*. 'I never shall.'[69]

* * * * *

In the late nineteenth-century, the wealth and riches promised by the Klondike Gold Rush gave many entrepreneurial minded women the opportunity to make a fortune of their own – if they were prepared to endure camping in blizzard conditions and the 3000-foot icy Chilkoot Pass to get it.

Born in Chicago in 1866, Martha Munger had an inauspicious start to life. 'Susan, I am disappointed,' her father remarked to her sixteen-year-old mother on looking at his newborn twin girls. 'I expected a boy.' Martha later wrote that when her mother relayed this story to her, she remarked, 'If my husband had said that to me I would have thrown those two babies at him.'[70] The Great Chicago Fire of October 1871 destroyed her father's laundry business, but his fighting spirit enabled him to re-establish himself and the Mungers were able to scale the Chicago social ladder – so much so that Martha was installed in the Lake Forest Select Seminary for Young Ladies to be 'finished.' The experience was difficult for all involved, but most of all for Martha who even in youth showed something of the adventurer she would become. Of those years, Martha later recalled that, 'I think I nearly finished those in charge of the 'Select Seminary.' I was continually getting into escapades. I was not disobedient, but my troubles arose from my zest for adventure. When seized by the urge of a particular quest, in pursuit of it I forgot all rules and regulations, thereby upsetting the discipline of the school. My teachers complained so often that finally my parents became too distraught to defend me and I was taken out.'[71] She was sent to a convent instead – where, surprisingly, she flourished.

Marriage quickly followed to the handsome Will Purdy, who had the entirely respectable and well-paid job of Paymaster on the Rock Island Railway. Then followed social events, parties, cycling clubs, Turkish baths and babies – all against the backdrop of the 'gay 90's.' Everything was starting to look and feel very conventional, until Martha and her family got word of the Klondike Gold Rush – which was inevitable given the numerous 'Gold! Gold! Gold!' attention grabbing headlines at the time. Gold was first discovered in the Klondike region of Yukon, north-west Canada in 1896 but its remote and mountainous location, on a river that froze for several months over winter, meant that the resulting stampede of prospectors didn't reach its peak until the summer of 1897. Between 1897 to 1898 it was estimated that up to 100,000 people tried to reach the Klondike gold fields to make their fortune, although two thirds of this number inevitably failed to reach their desired destination, and those that did often left empty-handed.

Back in Chicago, Martha was – in her own words – unhappy. She was ten years into her marriage and had produced two healthy sons. However, her husband was away much of the time due to his work on the railroad and boredom had set in. Martha was tired of the Turkish baths and cycling, and Will was tired of his job. Gold-prospecting was the answer to all their woes, and the couple, along with their friends, Mr and Mrs Gage, Martha's brother George, and her cousin Harry, made plans to travel to the Klondike. 'It looked like a great adventure,' Martha writes, 'and I was consumed with the urge to have my part in it.'[72] Martha later reflected on how she, and many others, had been sucked into the promise of limitless wealth,

> To me it was a quest that all the allure of a 'Treasure Island' or 'Aladdin's Lamp'…I pictured myself and children living in luxury the rest of our days. I look back now and wonder how anyone could have been taken in, as I was; could have been persuaded to go on such a wild goose chase; and that men with years of business experience – a railway president and a great banker – could have thought that such a huge fortune was to be had merely for the taking.[73]

But before they had even left Seattle, Will was called to San Francisco on business and the couple had to part at short notice. A week or so later, he wrote to Martha expressing his doubts about their proposed trip to the Klondike, particularly the challenges of getting there, and decided he'd changed his mind. Would she like to go to the Sandwich Islands instead? 'Go to the Sandwich Islands?' Martha exclaims. 'With my Klondyke ticket bought, my passage booked, my vision of a million dollars in gold dust? Even after 10 years of married life, how little Will Purdy knew me!'[74]

With total faith in her conviction that the Klondike held the key to eternal wealth and happiness, and absolutely no faith in her husband, Martha wrote to Will telling him that she had made up her mind to go to the Klondike as originally planned and that she would never go back to him, 'so undependable he had proven, that I never wanted to hear from or see him again.'[75]

Despite her conviction that she could 'go it alone' without the unreliable Will in tow, Martha had seriously underestimated the challenges that lay ahead. She set out from Seattle with her brother to steam the 1000 miles to Skagway, from where they would need to travel over the mountain ranges, including the notorious Chilkoot Pass (a forty-two-mile walk over a 3000-foot alpine trail) and down to Lake Lindeman. From there they would need to skirt the lake, pass down the rapids, and (hopefully) end up in one piece in the new 'town' of Dawson. 'With staff in hand, at last I had taken my place in that continuous line of pushing humans and straining animals,'[76] Martha wrote, although she soon regretted her bulky choice of attire – a silk and buckram lined corduroy-velvet skirt, pleated jacket, stiff collared blouse and 'voluminous brown silk bloomers'[77] – all of which impeded her ability to scramble over rocks and boulders with any great haste. But at least she was still alive. Their route was littered with the bones of dead horses and tumble-down shacks that had been abandoned – evidence of how unprepared many of the stampeders really were.

By the time they reached the foot of the Chilkoot Pass, news had reached them of the death of one hundred gold seekers in an avalanche, with more bodies being unearthed daily. 'I looked up the Pass,' Martha wrote in her memoir fifty years later. 'I can see it yet – that upward trail, outlined on an almost perpendicular wall of ice-covered rock, alive with clinging human beings and animals, slowly mounting, single file, to the summit.'[78] Undeterred, they joined the slow march of men, horses, dogs and native packers – all doggedly pursuing the same ill-fated plan as Martha. The journey was perilous, and an icy death leered at them from below from every precipice. Having negotiated narrow ledges, the width of her foot and bridges of ice over frozen streams, within ten feet of the summit Martha slipped and fell into a crevice. The accident earned her short shrift from her brother. 'For God's sake, Polly [George's pet name for his sister], buck up and be a man! Have some style and move on!' he said with exasperation as Martha sat weeping in the snow.

Worse was to come. The final two miles to Lake Lindeman were so awful that Martha begged to be left where she fell, but her brother and their agent managed to half-carry, half-drag her the rest of the way and install her in some lodgings. After a respite of three weeks, they packed up a boat and embarked on a terrifying journey down the White Horse Rapids to the tent-town of Whitehorse, where they were greeted by a plague of mosquitoes and spent an uncomfortable night. After twelve days of smooth sailing on the Yukon, they eventually reached their destination, Dawson – a ramshackle, shanty town on the make, where they could look forward to either a dose of malaria or typhoid fever, or a whisky at the 'Bucket of Blood' saloon - both equally unappealing and no doubt leading to certain death.

But danger was a little closer to home for Martha. She had belatedly realized that she was pregnant; a situation she would have to face alone and with little money. Worse, winter was setting in and she was now trapped. She couldn't risk the journey back over the Chilkoot and make her way back to Chicago, so at this point, death seemed preferable. Death was also a real possibility, given the circumstances in which the baby would have to be born. In fact, her cousin Harry, who travelled back to America that winter, told her relatives on his return that she was probably already dead.

Fortunately, Martha was made of stronger stuff. Like many women of her ilk when faced with a dreadful situation, she tackled it head on. She was determined not to be brought low by this newest, and potentially catastrophic development. So, she made baby clothes out of table napkins and pieces of fabric donated by other pioneers in the camp. The baby came early in January 1899, and she birthed her 9lb son alone in her cabin in the aptly named 'Lousetown,' although Martha reassuringly recalls that 'it was over quickly.'[79] She found the time to explore and take pleasure in her surroundings; admiring the swirl of the northern lights in the endlessly dark days and roaming the hills in search of orchids and wildflowers.

After surviving another fire – this time the Dawson fire of April 1899 – and a landslide, which obliterated the two cabins closest to Martha's but miraculously left hers intact, her father arrived to take her back to the safety of her family home in July of the same year. Yet despite her hardships, it wasn't what Martha wanted. 'What I wanted was not shelter and safety, but liberty and opportunity,'[80] she stated unequivocally. She would only accompany him back on one condition – that she be allowed to leave the gold claims she had staked at Excelsior Creek under the care of her brother, and should they fail to yield at least $10,000 before the following year she would never mention 'Klondyke' again. Back at her parent's ranch she quickly lost interest in life, and like many exploring women before her, found that the lack of an outlet for her sense of adventure had a profound effect on her physical and mental wellbeing. For a time, she became depressed and barely ate, until she was able to convince herself she had much to continue living for. And the Klondyke was still there in her thoughts;

> That vast new rugged country, its stark and splendid mountains, its lordly Yukon River, with all its streams and deep blue lakes, its midnight sun, its gold and green of summer, it's never-ending dark of winter, illumined by golden stars and flaming northern lights.[81]

To Martha's delight, the Klondyke came through – her investments paid off and she wasted no time in making the journey back to Dawson. She employed a crew of men to mine for her, and in 1901 set up a sawmill and quartz mill on the banks of the Klondyke, which made her plenty of money. Within a decade Dawson had become almost civilized, and Martha its First Lady. She married George Black, a solicitor with political ambitions in 1904 and earned her 'explorer stripes' in 1917 by becoming a Fellow of the Royal Geographical Society for her British lecture tour on life in the Yukon. A more illustrious position was to follow, when in 1935, at the age of almost seventy she was elected to the Canadian House of Commons, only the second woman ever to do so. She died on 31 October 1957 at the age of ninety-one in Whitehorse, the capital of her beloved Yukon – a life full of both adventure and misadventure well lived.

Part 5

For Duty

'If these sketches should prove the means of deterring one family from sinking their property, and shipwrecking all their hopes, by going to reside in the backwoods of Canada, I shall consider myself amply repaid for revealing the secrets of the prison-house, and feel that I have not toiled and suffered in the wilderness in vain.'[1]

Susanna Moodie, 1871

Susanna Moodie was a reluctant traveller. Having been dragged halfway across the world to live in an Upper Canadian wilderness populated by wolves, bears and unruly neighbours (both human and animal) her first impression of her new home was that it was distinctly mediocre. Despite assurances from their Yankee driver that it was in a 'smart location,' Susanna was far from convinced.

'You must be mistaken,' she said as she looked at the hovel in dismay, 'that is not a house, but a cattle-shed or pig-sty.'[2] On hearing that the miserable hut was indeed their new home, Susanna wept. To add insult to injury, the 'house' already had some inhabitants – three oxen and two cows, which 'were quietly reposing upon the floor.'[3] After the beasts had been reluctantly driven out, Susanna was left with her small baby, and Hannah the maid, alone in the 'untenable tenement,' surrounded on all sides by the dark forest, waiting for Susanna's husband to arrive.

It was vastly different from the 'land of opportunity' that was promised. Susanna and her family, along with approximately fifty thousand other hopefuls made up the first wave of British settlers to emigrate to Canada in the 1830s. In most cases, this was at the bidding of a husband or father whose interest had been piqued by the Government's pledge to provide land for farming, and tales of fortunes to be made. Women were summarily dragged along because that was what was expected of them, whether they wanted to go or not. This was a trial that was to be endured out of duty – and a trial it most certainly was. If one could come to terms with eating squirrels – or if times were really bad, family pets – sharing belongings with just about everyone in a three-mile radius; having said belongings 'borrowed' by numerous friends and neighbours (never to be seen again), sub-zero temperatures, and living under the constant threat of one's log cabin burning down in order to keep it warm, then apparently there were riches aplenty to be had.

Born in Suffolk, England, in 1803 Susanna was ill-prepared for her new life as a pioneer wife. In 1831, she married John Moodie, son of Major Moodie of Mellsetter in the Orkney Islands. As the youngest of three sons – a 'spare' rather than the 'heir' – John was expected to make his own way in life with none of the financial handouts gifted to his older brother. So, he decided to go to Lake Ontario in Upper Canada (now part of modern-day Ontario) and try to make his fortune there. And he most certainly wasn't the only person with the same idea. The press at the time dedicated a large chunk of their column inches to 'good news stories' from the Canadian colonies, printing letters from some of the first settlers with assurances that 'they are comfortably settled and in a thriving way.'[4] 'There are no robbers,' wrote one enthusiastic settler, 'and I have never yet seen a beggar; there is scarcely such a thing as a door with a lock, and many a house without even a door to close.'[5]

This was certainly true for Susanna. Her hut had no door because it was so dilapidated it had fallen off and someone had slung it into the back yard. It was finally retrieved and fitted back on, which made 'a great addition to our comfort.'[6] As for robbers and beggars, Susanna also learnt very quickly that the custom of 'borrowing' in Upper Canada meant something quite different to back home. Her first experience of her new neighbours was the appearance of 'a girl of seventeen or eighteen years of age, with sharp, knowing-looking features, a forward, impudent carriage, and a pert, flippant voice, standing upon one of the trunks, and surveying all our proceedings in the most impertinent manner.'[7] The girl proceeded to present Susanna with an empty whiskey decanter, allegedly at the request of her father who wanted them to borrow it, in case they didn't have their own. After the girl leaves, the servant James shrewdly comments that, 'That bottle is not brought here for nought.'[8] Yet Susanna, at this point still the trusting gentlewoman (before she had her peace of mind shattered by living in the middle of nowhere) 'could not unravel the mystery and thought no more about it.'[9]

The 'strange visitor' who was 'dressed in ragged petticoats...with uncombed elf-locks, and a face and hands that looked as if they had been unwashed for a month'[10] appears again the next day, and all becomes clear:

> 'Have you done with that 'ere decanter I brought across yesterday?'
>
> 'Oh, yes! I have no occasion for it.' I rose, took it from the shelf, and placed it in her hand.
>
> 'I guess you won't return it empty; that would be mean, father says. He wants it filled with whiskey.'[11]

Thinking this is a strange custom in her new country, Susanna fills up the decanter with some rum, hoping that in exchange she may get some milk for her daughter - only to find the rascally young lady wants payment for it. Susanna records that,

> Day after day I was tormented by this importunate creature, she borrowed of me tea, sugar, candles, starch, blueing, irons, pots, bowls – in short, every article in common domestic use – while it was with the utmost difficulty, we could get them returned. Articles of food, such as tea and sugar, or of convenience, like candles, starch, and soap, she

never dreamed of being required at her hands. This method of living upon their neighbours is a most convenient one to unprincipled people, as it does not involve the penalty of stealing; and they can keep the goods without the unpleasant necessity of returning them, or feeling the moral obligation of being grateful for their use.'[12]

The girl's father, who according to Susanna 'went by the familiar and unenviable title of Old Satan,' borrowed the Moodie's brand new plough and sent it back broken, as well as a spade and trowel which was loaned for an hour and never returned.

This was Susanna's induction into the Canadian 'borrowing' system, which seemed to amount to a practice that was little better than stealing, but everyone seemed to blindly accept as part of everyday life on the frontier. 'Nothing surprised me more than the extent to which this pernicious custom was carried, both by the native Canadians, the European settlers, and the lower order of Americans,' Susanna wrote. 'Many of the latter had spied out the goodness of the land, and borrowed various portions of it, without so much as asking leave of the absentee owners. Unfortunately, our new home was surrounded by these odious squatters…[13]'Little wonder the letters printed in the British newspapers about life in Canada urged potential settlers to 'bring all your crockery with you.'[14] The freeloading locals must have been running low on bowls and plates.

The shameful 'borrowing' practices of the locals were just one of many burdens Susanna had to bear in her new life. If she wasn't being battered into submission by the elements, she was attempting to tame them, or at least stop them from ruining her hard work and that of her husband – who managed to absent himself frequently at the worst possible times, leaving Susanna to cope with their smallholding and small children on her own.

In the winter of 1837, by which time the Moodie's had moved north to Kutchawanook Lake, temperatures dropped as far as thirty degrees below zero, which was as far as the line on Susanna's thermometer went. Everything that was liquid froze and in the ensuing chaos of attempting to feed and warm up her four long-suffering children, Susanna didn't notice that the newly appointed Irish maid had stuffed the Franklin stove with cedar chips, which very quickly overheated all the way up to its topmost pipe which let out smoke through the roof. Having dealt with the overheated stove, she sent the maid out to check the roof. 'She quickly returned,' recalled Susanna, 'stamping and tearing her hair, and making a variety of uncouth outcries, from which I gathered that the roof was in flames. This was terrible news...'[15]

It most certainly was since John was on one of his 'trips' and they were a mile and a quarter from the nearest habitation.

Susanna dispatched the 'weeping and lamenting' maid to get help, who shrieked "Fire!' the whole way. This was utterly useless, and only impeded her progress by exhausting her strength,' Susanna bemoaned. She was left, with her children, 'quite alone, with the house burning over my head.' She had two choices – neither of which were particularly appealing. Either stay in the house and be burnt to death or leave the house and freeze to death. She chose the latter, naturally ensuring the children's safety before her own by protecting them from the cold. Susanna, 'emptied all the clothes out of a large, deep chest of drawers, and dragged the empty drawers up the hill; these I lined with blankets, and placed a child in each drawer, covering it well over with the bedding, giving to little Agnes [the Moodie's youngest daughter] the charge of the baby to hold between her knees, and keep well covered until help should arrive.'[16] At her elder daughter Katie's insistence, Susanna went back inside the burning cabin to save her husband's flute (as any mother will testify, the thought of risking her life for the flute was probably a more appealing prospect than listening to her daughter whine about it getting lost in the fire for hours on end) and shortly afterwards, John Moodie came running towards her at full speed with some other men and managed to dampen the fire before it destroyed the walls of the house. The only casualty was the family dog, Snarleyowe. The flute escaped the ordeal unscathed.

As the genteel youngest daughter of a family of six girls who by the time of her emigration in 1831 was already a published author of poetry and children's stories she was thoroughly unsuited to a life of roughing it in the bush. John Moodie's credentials were only marginally better suited to farming: he was a soldier by trade, serving in the 21st North Royal British Fusiliers before being seriously wounded in Holland in 1814 during the Napoleonic Wars. He had also been a magistrate in South Africa, until in 1829 he took up his pen and became a writer like his future wife. To fulfil their farming ambitions the Moodie's had swapped their comfortable English town house for a shed in the woods and drawing room poetry readings for arguments with the local riffraff about ownership of a kettle. Susanna was stoic though, writing that 'Emigration, in most cases – and ours was no exception to the general rule – is a matter of necessity, not of choice. It may, indeed, generally be regarded as an act of duty performed at the expense of personal enjoyment.'[17] And there was certainly no enjoyment to be had in the gloomy forests on the banks of Lake Ontario. Susanna's ordeal lasted seven years and she hated every minute of it.

At the same time as Susanna was putting out fires and bickering with the locals, her older sister Catharine Parr Traill, who was born in 1802, was treading a similar path a mile and a half down the road. Catharine arrived in Upper Canada in 1832, along with her husband Thomas Traill. Like Susanna, she too was a writer with an impressive back-catalogue, and her experiences are well documented in the ludicrously long-titled *The Backwoods of Canada; being letters from the Wife of an Emigrant Officer, Illustrative of the Domestic Economy of British America* (1836). But unlike Susanna, Catharine accepted her lot and even managed to embrace it. In fact, instead of complaining about the shenanigans of the locals, she set her mind to writing a pragmatic account of life in Upper Canada, free of the opinions and emotional drama that made Susanna's account such an entertaining read. In this respect, Catharine more than fulfils the expectation to 'do her duty' – she is the model pioneer wife, despite facing similar hardships as her sister; fitting neatly into the cast iron framework of Victorian society. Her practical approach to adversity meant that where Susanna struggled to revise her expectations of what life in the bush would be like - and accept it was a backwards step - Catharine was able to lower hers sufficiently to adapt to her new environment and remain stoically focused on the future – seeing possibility in the untamed landscape rather than doubt and despondency. Susanna's fatalistic outlook was common – the shock of the new, and the fact that many emigrants had been lured to Canada with the promise of lands and livelihoods aplenty, meant that hopes were raised and then dashed as soon as they realized the extent of the work ahead.

But instead of complaining, Catharine threw herself into it and got her hands dirty with the best of them. She was there now, and she meant to stay regardless of cholera (which she caught), fierce weather (of which there was much), birthing several children (nine in total) and the fact no one knew how to run a farm (which meant they had to muddle through and hope for the best). All this she achieved and then she very helpfully wrote about it so no one would be lulled into thinking that life in Upper Canada was anything less than an ordeal.

Fortunately, Catharine was given a reality check before she so much as stepped on a forest twig, let alone cut down a tree. Her bubble was burst by a fellow traveller who had emigrated some years earlier. On hearing Catharine's plans to make their log cabin presentable he reliably informed her (no doubt with a certain amount of schadenfreude) that they would have so many obstacles to encounter in the first two or three years 'that you will hardly have opportunity for carrying these improvements into effect.'[18] He then predicted that 'at the end of ten or fifteen years you may begin to talk of these pretty improvements and elegancies, and you will then be able to see a little what you are about.'[19]

She replied that 'all the works on emigration that I have read… give a fair and flattering picture of a settler's life; for, according to their statements, the difficulties are easily removed.'[20] Her fellow traveller told her to ignore the books and 'use your own reason,' which was probably the best piece of advice she received during her time in Upper Canada. 'I began to apprehend that we also had taken too flattering a view of a settler's life as it must be in the backwoods,' she wrote. 'Time and our own personal knowledge will be the surest test, and to that we must bow.'[21]

Catharine's 'tests' included losing the oxen they were using to drag the logs they needed to build their cabin (such was the oxen's dislike of the bush they walked twenty miles back to their previous owner); frozen plaster - meaning they couldn't finish decorating their cabin; intermittent mosquito assaults; hurricanes and snowstorms; crop failure; financial struggles; tornados; snow blindness; thieving squirrels; and the ever-present threat of house-fire. She coped with all of this on a diet of potatoes and 'Yankee tea' (boiled hemlock sprigs) which was 'vile.'[22]

But she faced her challenges with unswerving cheerfulness and equanimity. 'We begin to get reconciled to our Robinson Crusoe sort of life,' she notes, 'and the consideration that the present evils are but temporary goes a great way towards reconciling us to them.'[23]

Catharine is clear who *would* and who certainly *would not* make a good bush-settler. 'The poor hardworking sober labourers, who have industrious habits, a large family to provide for, and a laudable horror of the workhouse and parish over-seers,' she suggests. 'This will bear them through the hardships and privations of a first settlement in the backwoods.' To succeed, Catharine notes the following qualities must be in evidence: 'perseverance, patience, industry, ingenuity, moderation, self-denial...'[24] If you didn't have nineteenth-century grit in abundance then you need not apply. For Catharine this includes, 'The poor gentleman of delicate and refined habits' who is 'unwilling or incapable of working himself.' These poor saps are 'not fitted for Canada, especially if his habits are expensive.'[25]

The same applies to the ladies. The 'wives and families... accustomed to the daily enjoyment of every luxury that money could procure or fashion invent; whose ideas of happiness are connected with a round of amusement, company, and all the novelties of dress and pleasure that the gay world can offer.' More specifically, 'Young ladies who have been brought up at fashionable boarding-schools, with a contempt of everything useful or economical.' Such people would make 'very indifferent settlers' wives.'[26]

For there was no fun to be had in the bush. Just work, work and more work – although Catharine herself found some diversions, deriving pleasure from her explorations of the landscape. She dedicated numerous chapters of *The Backwoods of Canada* to her study of the natural world around her, delighting in the new and strange (to her anyway) species of birds, flora and fauna on the lake and in the woods. The red-headed woodpeckers were 'splendid,' the titmouse had a 'gay warbling,' and the humming-birds were 'beautiful.' She wasn't overly keen on the blackbirds though, as they were apparently too 'saucy.' She was more enraptured by the plants and flowers, waxing lyrical over the 'magnificent water-lily,' that she named 'Queen of the Lakes, for she sits a crown upon the waters…The shallows of our lakes abound with a variety of elegant aquatic plants: I know not a more lovely sight than one of these floating gardens. Here you shall behold near the shore a bed of azure fleur-de-lis, from the palest pearl colour varying to the darkest purple…'[27] and so on and so forth.

Her descriptions are rich with a full palette of colour, textures and scents and one wonders how she found the time for such in-depth study amongst all her labouring. Despite her best efforts to give a 'warts and all' account of life in the bush, she still manages to paint an idyllic picture. Even the frogs were 'handsome.'

She also managed to find joy in the inhospitable elements, describing the frost and near-arctic conditions in similar raptures to the flowers:

> The earth, the trees, every stick, dried leaf, and stone in my path, was glittering with mimic diamonds, as if touched by some magical objects the most rude and devoid of beauty had suddenly assumed a brilliancy that was dazzling beyond the most vivid fancy to conceive; every frozen particle sent forth rays of bright light. You might have imagined yourself in Sinbad's valley of gems; nor was the temperature of the air at all unpleasantly cold.[28]

Her sister Susanna, had rather a different experience:

> The winter had now fairly set in – the iron winter of 1833. The snow was unusually deep, and it being our first winter in Canada, and passed in such a miserable dwelling, we felt it very severely. In spite of all my boasted fortitude – and I think my powers of endurance have been tried to the utmost since my sojourn in this country – the rigour of the climate subdued my proud, independent English spirit, and I actually shamed my womanhood, and cried with the cold.[29]

Even Susanna's choice of language is telling, or at the least hopeful. Her 'sojourn' lasted a lifetime. Susanna died in Toronto, Ontario on 8 April 1885, having spent the rest of her life in Canada. In the introduction to the 1871 edition of *Roughing it in the Bush* (which was first printed in 1852) Susanna reflected on her journey as an emigrant's wife. She admits that it was written as a, 'warning to others...not to take up grants and pitch their tents in the wilderness, and by doing so, reduce themselves and their families to hopeless poverty.'[30] But she conceded that she had come to see Canada as home, writing that 'my love for the country has steadily increased, from year to year, and my attachment to Canada is now so strong, that I cannot imagine any inducement, short of absolute necessity, which could induce me to leave the colony, where, as a wife and mother, some of the happiest years of my life have been spent.' Praise indeed! Susanna continues, 'Contrasting the first years of my life in the bush, with Canada as she now is, my mind is filled with wonder and gratitude at the rapid strides she has made towards the fulfilment of a great and glorious destiny.'

Forty years after putting her first step on Canadian soil, Susanna could claim the accolade of being a true pioneer. For her, Upper Canada's progress could be measured in its cultural, artistic and scientific advancements. 'Institutes and literary associations for the encouragement of learning are now to be found in all the cities and large towns in the Dominion,' she wrote proudly. There were publishing houses in Toronto and Montreal and 'no lack of native talent or books, or of intelligent readers to appreciate them.' Works of art depicting 'the grand scenery of our lakes and forests...adorns the homes of our wealthy.' There were civic buildings; numerous fine examples of engineering and architecture; and welfare provision for the less fortunate. Upper Canada was no longer an untamed wilderness. Arguments about who last borrowed the village kettle and why it hadn't been returned were a thing of the past. Now everyone had their own kettle! It was *civilised*.

Catharine decided to dedicate a whole book to putting other people off emigrating to Canada. *The Female Emigrant's Guide and Hints on Canadian Housekeeping* (1854) was a sensibly written (read: *candid*) handbook 'for wives and daughters of the future settler, who for the most part, possess but a very vague idea of the particular duties which they are destined to undertake, and are often totally unprepared to meet the emergencies of their new mode of life.'[31]

Catharine meant well – her aim was to mitigate the sense of uncertainty and lack of knowledge experienced by the female émigré. If Catharine had a mantra, it was almost certainly *forewarned is forearmed*, and she did her utmost to fill in the gaps that she herself had had to fill in through bitter personal experience. But that also meant explaining what that experience entailed, which made for a depressing preface.

> ...females have everything to learn, with few opportunities of acquiring the requisite knowledge, which is often obtained under circumstances, and in situations the most discouraging; while their hearts are yet filled with natural yearnings after the land of their birth, (dear even to the poorest emigrant), with grief for the friends of their early days, and while every object in this new country is strange to them. Disheartened by repeated failures,

unused to the expedients which the older inhabitants adopt in any case of difficulty, repining and disgust take the place of cheerful activity; troubles increase, and the power to overcome them decreases; domestic happiness disappears. The woman toils on heart-sick and pining for the home she left behind her. The husband reproaches his broken-hearted partner, and both blame the Colony for the failure of the individual.[32]

If a potential immigrant managed to read past Catharine's disparaging words, they could anticipate lots of helpful advice - from decorating ('It is a great mistake to neglect those little household adornments which will give a look of cheerfulness to the very humblest home'[33]), to furniture ('A set of corner shelves, fitted into the angles of the room, one above the other, diminishing in size, form a useful receptacle for any little ornamental matters'[34]) to gardening ('The commonest climber for a log-house is the hop…Planted near the pillars of your verandah, it forms a graceful drapery of leaves and flowers, which are pleasing to look upon'[35]). Catharine was keen to impart some wise words to her gentleman readers too. 'Whatever be the determination of the intended emigrant,' she wrote, 'let him not exclude from his entire confidence the wife of his bosom, the natural sharer of his fortunes, be the path which leads to them rough or smooth. She ought not to be dragged as an unwilling sacrifice at the shrine of duty from home, kindred and friends, without her full consent.'[36]

Catharine's guide must have proved popular as it was reprinted several times as the *Canadian Settler's Guide*.

Catharine was well-suited to a rough life. She outlived all of her siblings - who were themselves long-lived – her husband and, sadly, several of her own children. Born in 1802, her life spanned almost the entire nineteenth-century before she died at her cottage 'Westlove' in Lakefield, Ontario in 1899, at the age of ninety-seven. Her greatest contribution to Canada was her early record of its colonization and her books are amongst some of the very earliest in the Canadian literary canon. 'The greatest heroine in life,' she wrote, 'is she who knowing her duty, resolves not only to do it, but to do it to the best of her abilities, with heart and mind bent upon the work.'[37]

Catharine died safe in the knowledge that she'd followed her own advice to the letter.

Catharine Traill and Susanna Moodie may have been settlers rather than explorers, but their circumstances allowed them to discover the limits of their resources and the strength of their resolve. As they became accustomed to their new and strange external landscape, internally they were also going on a journey of self-discovery, mapping out a future for themselves and their families. They trod a wild and unexplored path and by doing so, smoothed the way for others.

Eliza Bradley's story was shorter than Catharine and Susanna's but more catastrophic. For such a short tale of woe, it must have the longest title ever conceived: *An Authentic Narrative of the Shipwreck and Sufferings of Mrs Eliza Bradley, The Wife of Capt. James Bradley, of Liverpool, Commander of the Ship Sally, which was Wrecked on the Coast of Barbary, in June, 1818.* Eliza's account begins in Liverpool, where she was born in 1783 and married in 1802 to Capt. James Bradley, who was 'bred to the seas.'[38] Despite Eliza's best efforts to 'induce him to pursue some other occupation,'[39] her endeavours proved completely ineffectual, so they packed up their belongings and set sail for Tenerife on the *Sally* – the ship which Capt. Bradley had been appointed to command – on 18 May 1818. Eliza was the only female amongst the thirty-two souls on board. All was going to plan until five weeks into their journey when they were caught up in a furious storm for six days - which must have been an utterly terrifying experience in the pre-ocean liner age. The ship started leaking in various places and 'it was soon concluded by the officers utterly impossible to save either the ship or their effects; the preservation of even their lives becoming every moment more difficult to them,' Eliza recounts, '[and] they now began to apply every thought and deed to that consideration.'[40]

The ship, which was now considered 'unmanageable,' had been totally blown off course and nobody seemed to know where they were. The crew thought they might be near to land, but no one could say so with any certainty – until they struck a chain of rocks, which completely tore open the stern. All hope was lost and 'the ship resounded with their [the passengers and crew] lamentable exclamations, imploring the mercy of their Creator!'[41] Eliza claims she kept her head throughout the ensuing melee (which one would like to believe is true but sounds a little far-fetched given the circumstances). She writes that 'the extremity of the misfortune, with the certainty of its being inevitable, served to supply me with a sort of seeming firmness.'[42]

They spent several days clinging on to the wreck – for dear life – and drifting aimlessly, whilst all the while being tossed and thrown about the sea; the waves crashing 'like mountains,' over their heads, and being pelted with rain, which was falling in torrents.

Finally, someone spotted land, but all agreed it would be suicidal to attempt to swim the distance to the shore. On realizing this, 'the whole crew was seized with the extremity of despair; their groans and exclamations redoubled and were repeated with such strength and fervency, that they were to be heard amidst the raging of the winds, the roaring of the thunder, and the dashing of the waves, which, all joined together, augmented the horror of the scene.'[43]

They spent the night clinging to the stricken ship, torturously close to land but with no means of reaching it. Death would have been a deliverance at this point, as far as the crew was concerned, but Eliza had reserves – she was fighting for her life. It was quite literally a sink or swim situation. They passed a day and a night in this sorry predicament before someone had a bright idea. A small boat attached to the main vessel, that for some bizarre reason no one had thought to use before, was elected to carry at least *some* of the crew to safety. But for obvious reasons, no one was particularly keen on being in the second cohort of people to be rescued, but Eliza smugly writes that, 'It was, however, unanimously agreed by all, that my husband

and myself should be among the number who should go first into the boat.'[44] Lucky them.

The entire crew, along with Eliza and Captain Bradley, did eventually reach land and a search party managed to salvage some timbers for a shelter, a barrel of flour and a keg of salted pork. Captain Bradley ascertained by their position that they had washed up on the Barbary coast of north-west Africa. That no one was lost at sea during their ordeal, given the ferocity of the waves as described by Eliza is a little circumspect – particularly as many nineteenth-century sailors couldn't swim. But since Eliza doesn't mention any lost souls, we can only assume the whole crew survived – either that or Eliza was prone to a little embellishment.

The horrific ordeal on the ocean was only the first in a catalogue of disasters. They managed to negotiate the rocky cliffs they landed on, only to find a vast sandy plain stretching away into the horizon and no sign of civilization. At one point Eliza threw herself down on the sand and refused to move. She had resolved to die and that was that. In fact, she willed it upon herself because dying seemed preferable to traipsing across an unknown land trying to find help: 'I was indeed so determined to die, that I awaited the moment with impatience, as the termination of my misery.' Captain Bradley, by now no doubt regretting the idea to bring his wife along, tried his best to rally her spirits by suggesting that they might meet with friendly aid, 'by the means of which we might be conducted to some commercial port, at which we might probably obtain a passage for Europe.'[45]

What happened, according to Eliza, was the opposite. The crew marched on, foraging for some meagre sustenance along the way (wild sorrel, a dead seal and some mussels, washed down with stagnant pool water), whilst Eliza lamented her misfortunes, freely admitting that 'the miseries that I had endured since that melancholy event [the shipwreck], had afforded me but little leisure to reflect on the situation of any one but myself.'[46] The situation didn't improve when a search party looking for water was chased by a band of naked

Arabs, armed with muskets, spears and scimitars, who then followed the search party straight back to the rest of the hapless group.

All were captured, including Eliza, and stripped down to their underwear. The Arabs then started fighting amongst themselves to claim each crew member as their own property. Once this was settled, which according to Eliza took an hour and several nasty flesh wounds, they were given water and Eliza was permitted to ride on a camel in exchange for her 'gown, bonnet, shoes and stockings,' leaving her dressed only in her 'petticoat and shimmy.' The caravan then spent several days trekking across the desert in unrelenting heat, only stopping to pitch their tents for the night or to source water and sustenance. 'Having been deprived of my bonnet, and having nothing to defend my head from the sun's scorching rays, the pain that I endured was extremely excruciating,' wrote Eliza of her inevitable sunburn. The whole party suffered the ill-effects of the sun, which 'scorched and blistered' their bodies from head to foot. But worse was to come for Eliza.

She was permitted to spend several minutes a day in conversation with her husband, and during one of these exchanges she learnt that they were to be separated. Captain Bradley had discovered from his 'master' (the Arab who had claimed him as his possession) that the caravan would separate. He was bound for Mogador (now the port of Essaouira), where the Arabs expected to fetch a decent price for his life at the English Consul. Since she belonged to a different 'master' who refused to part with her (as a female captive she was expected to fetch double the price at the English consul) and who wasn't planning to go to the consul yet, her journey from the point at which they were separated was anyone's guess.

After more tedious trekking, punctuated only by meals of dubious quality (roasted snails, dead camels, goat entrails) and water-stops, Eliza arrived at what would turn out to be her home for the next five months. The captives were welcomed heartily by the locals with actions rather than words: 'spitting on us, pelting us with stones, and throwing sand in our faces, accompanied with the word, 'fonta' (bad.)'

Eliza recounted later. Here, she would have her final conversation with her husband before they were separated:

> It would be impossible for me to describe my feelings at this moment, and the reader can have but a faint conception of them! I begged that I might be indulged with the liberty of exchanging a few words with my husband, previous to his departure; but even this privilege was denied me. In a fit of despair I threw myself upon a mat, where I remained in a state of insensibility until the captives were far out of sight.[47]

With Captain Bradley gone, Eliza had only her own fortitude to sustain her. Her plight was met with little sympathy from the local women and children, who would regularly visit her for the sheer amusement of spitting in her face and throwing sand on her. Surprisingly, her Arab master attempted to make her circumstances as comfortable as possible – she was allowed to have a bible, was regularly fed and given water, and permitted to wander around the village. When he discovered how the villagers had treated her, he was furious. 'On his arrival, viewing the sad condition that I was in, with my eyes and mouth filled with sand,' Eliza recounted after one such incident, 'he became greatly enraged, and beat the vile authors of it unmercifully.'[48] Whether this was due to genuine concern for her welfare or simply because he knew that to deliver her up to the British Consul half-dead and traumatized would not bode well for him, it is difficult to say, but Eliza's dire circumstances were much relieved by his attentiveness. He also made it clear that if she tried to run away from him, her punishment would be instant death – just in case Eliza was in any doubt as to who was still in charge.

Her period of captivity afforded Eliza a chance to reflect. With only the Bible for company, she turned to the Scriptures for comfort and guidance, which also gave her the chance to preach to 'ye fair ones of Britain,' in other words middle-class British women, like herself. Those who,

> doat on the parade of public assemblies, and sail along in the full, blown pride of fashionable attire, of which the least appendage or circumstances must not be discomposed; thoughtless of human wo;, insensible to the sad condition of those, like myself, pining in many a solitary residence of want; ye gaudy flutterers, 'with hard hearts under a soft raiment,' how much more brilliant, as well as beautiful, would ye appear in the eyes of saints and angels, were you to employ your leisure hours, thus devoted, to the attaining, a knowledge of that Sacred Scripture, by which alone ye can expect to attain eternal life.[49]

After this religious interlude she then turned ethnographer, describing the Arab villagers in some detail, including their complexion, hair, beards, clothes and customs. She was shocked by how they treated their wives ('Their husbands consider them as their inferiors, as beings without souls') and their choice of faith ('O, what a pity is it that they are not taught the superior excellence of the Christian religion…Weep, O my soul, over the forlorn state of the benighted heathen!') and their attempts at teaching her to weave, which judging by Eliza's comment that 'they finally gave up all hopes of rendering me serviceable to them in this way,' we can assume was unsuccessful.

Eliza's deliverance did eventually arrive in the form of a letter from her husband, notifying her that the consul, a man named William Willshire, had agreed to pay $700 for her release. After being overcome by tears, she gave thanks to God (naturally) and embarked on the seven-hundred-mile trip to Mogador. It took nearly twenty days (and yet more trekking through the desert) but she finally arrived in Mogador and was reunited with Captain Bradley. 'Happy meeting!' she writes, 'It was some time before my husband or myself could exchange a syllable with each other. The joy which we both felt in being enabled to meet again under circumstances so different from those under which we parted, deprived us for some time of the power of speech.'[50]

On 1 February 1819, almost nine months after she first walked up the gangplank of the doomed *Sally*, Eliza was back on board a boat alongside her husband. This time she was destined for her home in Liverpool, which she reached forty days later.

Eliza's account reads like a work of fiction. Indeed, the author of *Wayward Women*, Jane Robinson, likens the published version to a 'flimsy popular chapbook, with crude and primitive woodcuts and a good fruity flavour.'[51] The title claims it is 'An Authentic Narrative,' but is it? It is an intriguing question, particularly given the popularity of texts relating to piracy and captivity at that time, such as Captain James Riley's *Sufferings in Africa* (1817) and Mrs Maria Martin's (fictional) account, *Mrs Maria Martin Who Was Six Years a Slave in Algiers: Two of Which She Was Confined in a Dark and Dismal Dungeon, Loaded with Irons, for Refusing to Comply with the Brutal Request of a Turkish Officer* (1807).

The first of these was a very real account. Captain James Riley did indeed exist and was shipwrecked off the coast of what is now the Western Sahara in 1815. He was also captured and kept as a slave, along with his crew – an ordeal which led to his memoir, *Authentic Narrative of the Loss of the American Brig 'Commerce' by the 'Late Master and Supercargo' James Riley*, which at some point was re-titled to the much more succinct, *Sufferings in Africa*. Riley was so affected by his ordeal at the hands of his native captors that once home in America, he devoted himself to anti-slavery work. He also founded the midwestern village of Willshire, Ohio – so named because the British consul for Mogador who eventually secured Riley's release was no other than the same William Willshire referenced in Eliza Bradley's account.

The provenance of Eliza's narrative is less certain. What is striking are the parallels between her work and those that came before – according to critics of the narrative, Eliza 'borrows' much of her story from Captain Riley's, and in some cases directly copies from it. For instance, her description of her 'master' bears much resemblance to the Arabs described by Riley – even down to his 'red and fiery' eyes, which appears in both accounts.

Stories such as Eliza Bradley's and her real-life shipwrecked counterparts were popular in the nineteenth century. Readers craved these near-death experiences, where the hero or heroine – who was nearly always white and European or American - triumphs over nature and then his or her captors – who unsurprisingly were nearly always natives; the villainous 'other.'

The mystery doesn't end there either. Despite claiming to be an English woman (as does another unreliable narrator, Maria Martin), there is no proof that an English edition of Eliza Bradley's account ever existed. There certainly isn't one listed in the British Library catalogue, although they do hold three editions printed between 1821 and 1823 in Boston, and a further American edition printed in 1824 in Exeter, New Hampshire. The preface of the Boston edition boasts, 'This publication having passed through a number of editions in England, and received an uncommon degree of patronage there, being printed almost verbatim from the original manuscript of Mrs B. as the English publisher declares; we are induced to present it to the American public, under a full conviction of its being patronized in proportion to its merits.'[52] Just to muddy the water even further the publisher also declares that the account must be true because the 'European edition, from which this is copied, being acquainted with the family of the writer of this narrative, and the circumstances of the unfortunate voyage upon which it was founded, clearly demonstrates the truth of the facts herein contained.'[53] Her story never made it into the British newspapers either, which seems remarkable given how sensational it would have been at the time.

Bradley's account became so fixed in the imagination as an authentic narrative, that it became a true story not just for nineteenth-century readers but for their modern-day counterparts. We will never know the truth of it, but there is an interesting footnote to Eliza's story. In 1828 an American 'Travelling Museum of Fine Arts' advertised its forthcoming exhibition in Hagerstown, Maryland. Part of the exhibition was a variety of waxworks of historical figures, such as Napoleon Bonaparte, Robert Bruce, Joan of Arc, and the well-known British explorer and heroine, Eliza Bradley 'who was taken prisoner by the Arabs on the coast of Arabia.'

Even if Eliza Bradley was fictitious, she did at least exist in wax.

* * * * *

Mrs Cara Edgeworth David definitely existed. We know this because she was kind enough to print her photograph, along with her husband and scientist friends, at the front of her book *Funafuti, or Three Months on a Coral Island*. She also professes in her preface that, 'The leader of the expedition says that as my MS [manuscript] has thrice made him fall asleep, the book will probably prove an infallible cure for insomnia, and is worth publishing as such,' which seems a little harsh given the subject matter she was charged with enlivening was a coral-boring expedition, led by her husband. Cara's self-professed (and highly amusing) 'Unscientific Account of a Scientific Expedition,' was published in 1899, and she makes it abundantly clear from the first line that she is tagging along resignedly,

> To start with, we none of us knew where it [Funafuti] was; but my husband said he was going, and politely hinted that I should be an idiot not to go with him. After that there was nothing for me to say; he said it was a great opportunity that one might never get again. I hope we shan't; but then some people are never properly grateful for their advantages.[54]

Despite her husband's insistence that she accompany him, it didn't prevent him from telling her, fairly bluntly, when she was getting in the way – but like all good nineteenth-century wives (and with an inexhaustible amount of patience) she remains cheerful throughout, even when presented with her new and underwhelming accommodation, although by her own admission, biting her tongue nearly finishes her off. 'I had stupidly promised never to grumble during the expedition,' she writes, 'the self-repression thereby entailed nearly killed me, and so with outward cheerfulness I accepted the hut.'[55]

So, whilst Professor David and his five coral-fanatic friends busied themselves with their scientific experiments on the reefs, the uncomplaining and ever-resourceful Mrs David gets herself settled into island life. Her witty account details a run in with some terrible headwear in church ('It took us all sermon time to recover from the shock those hats had given us'[56]), the disappointment of a hastily built tool-shed ('To the left, looking even more hideous than any other corrugated iron building I had ever seen, was our store-shed. It was an insult to the lovely palm grove in which it stood'[57]), and several run-ins with local disease and illness, which she did her best to treat ('Another man tried to stop the sickness by drinking half a pint of kerosene oil, after which he felt so bad that he came to me for medicine; I felt inclined to prescribe a lighted match, but thought there might be trouble for me at the inquest'[58]).

When she wasn't attending to the medicinal needs of the local population, or doing her housework, she found much to do. She made notes on local folk songs and stories, learned about the island language and customs, joined in with community dances and singing, swam in lagoons and even visited and taught at the local school – where she discovered the girls were learning to make their own hats (which explained the awful headwear from earlier). But she keeps her sense of humour, which is the single most important trait that a wife-in-tow can have at her disposal if she is to last the distance. That, and the ability to hold one's tongue.

Unlike Cara David's coral expert husband, Mrs Isobel Gill's husband was more interested in what was above the ocean, rather than what was in it. Stars and the solar system were David Gill's passion – more specifically, his desire to solve 'a great problem, viz., *the distance of the Earth from the Sun.*'[59] His 'great problem' saw both of them travelling in 1878 to the isolated volcanic island of Ascension, in the South Atlantic, which at that point was little more than a collection of barren lava fields with only one green space – a man-made 'cloud forest' on the top of the highest mountain peak. If the reader of Isobel's narrative can get through its forty-six-page introduction, which her husband hijacks as a chance to impart a good deal of extremely scientific-sounding knowledge, we get a sense of the place where they were to spend the next six months.

'What a sight it was!' Isobel records on first seeing the island. 'The 'Abomination of Desolation'' where 'Stones, stones, everywhere stones, that have been tried in the fire and are now heaped about in dire confusion, or beaten into dust which we see dancing in pillars before the wind. Dust, sunshine, and cinders, and low yellow houses frizzling in it all! Is *that* Ascension?'[60]

It wasn't a good start. But Isobel adjusts quickly and proves herself a worthy assistant to her astronomer husband. They start by building an Observatory on the local croquet ground – if there was any doubt as to which Empire the island belongs - only to find that the location they have chosen is plagued by cloudy skies, which makes it impossible for them to take readings with their astronomy equipment and therefore record anything of note. 'This lovely, changing sky I could not love,' laments Isobel, 'for empty pages, where figures should have been, lay open by the Heliometer.'[61]

The Heliometer – the vital instrument they were using to take measurements – was bothersome even before they embarked on their trip. At a trial of the Heliometer at Burlington House for the benefit of the Royal Astronomical Society, ten days before the couple were due to leave, the instrument came crashing down, driving itself into the floor and smashing to pieces. Isobel records that, 'The apparent ruin of so many hopes and plans was paralysing, and for some minutes David was quite incapable of examining the amount of damage done.'[62] Fortunately, the endeavours of several opticians put it right before they were due to leave, but the Heliometer wasn't done with them yet. Having given up on their first location due to the irritations of the weather, Isobel and David decided to try their luck on the other side of the island, which gives Isobel the chance to stretch her explorer's legs. She volunteers to scope out a new spot - since David cannot leave his observation post - and after the usual declaration that it would be impossible for *a woman* to achieve such a thing ('You have never been beyond the Garrison,' David helpfully reminds her) Isobel gets the opportunity to show some grit…by traipsing across several miles of the stuff.

'At 10 P.M. we started,' Isobel writes. 'My spirits were higher than they had been for many nights.'[63] She now had a delicious – if short lived – taste of freedom. 'A sort of awe, not unpleasant, but the reverse, was stealing over me, and I felt just in the mood for an adventure; when, lo! close to my ear, a shrill uncanny shriek rang out through the stillness.'[64] Fortunately, it was only a one-eyed donkey, rather than the wild cats her husband had warned her of. Reunited with David and having given him a report of her findings (that those pesky clouds weren't so noticeable further south), he announces his intention to move all the equipment over the barren landscape forthwith. The only sticking point is the Heliometer and how to transport it without it breaking again. There is a slightly hairy moment when the Heliometer, which was lashed to a mast and carried on the shoulders of eight natives finds its way onto their heads instead – much to David's horror. What the native Kroomen thought of their white visitors and their strange star-gazing endeavours, one can only guess, but the Heliotrope did make it to its new location in the newly christened 'Mars Bay' in one piece, despite almost giving David Gill a nervous breakdown.

When they didn't have their eyes glued to the Heliometer, the Gills found time to explore the coastline, remarking on the birds and fish, and enjoying a spot of rock-pooling, hill-walking, expeditions to the 'cloud forest' on Green Mountain, as well as writing letters and reports home. It was a quiet life, but they had company in their

Kroomen assistants and a small menagerie of animals, including a poodle named Rover; a 'pathetic' but 'gentle-mannered' dog named Beauty; and Polly - 'a most uninteresting parrot…whose days were spent entirely in eating and screaming.'[65]

The inevitable worries over tea-making surfaced ('I had no tea that afternoon, so I remember it as a Black Letter Day, for even Ascension tea is better than none'[66]) as well as an inevitable host of disagreeable meals ("Eels!' I exclaimed in disgust; and threw down the fork which was conveying a choice morsel to my lips'[67]) but they did eventually come to love the dramatic and unforgiving landscape of Ascension. 'For the first time I saw beauty…Grim and joyless, but grand and majestic, were these gloomy rocks, trimmed round the base with delicately-tinted coral, their sternness veiled in feathery foam.'[68]

The trip was deemed thoroughly successful and shortly afterwards, David was appointed to the prestigious post of Her Majesty's Astronomer at the Cape of Good Hope, where the couple stayed for the next twenty-seven years. Whilst David spent his time in South Africa advancing the field of astronomy, Isobel set her mind to improving the lives of those around her. She established hospitals for the poor and created a scholarship for female students at the Cape University, amongst other good works and philanthropic ventures. The Gills retired to London from South Africa in 1907.

They were a formidable pairing. A photograph of the couple in their retirement shows them in a homely setting, sitting in two armchairs by the fireplace – David reading and Isobel knitting. At the time, David is said to have remarked, 'We are a very 'Darby and Joan' old couple, who like to be together as much as possible.' Which is extremely sweet - and very convenient for David who may have had a harder time of it had Isobel not been just as interested in astronomy and exploration as him. They were eventually united in death too – David departing first in 1914, and Isobel following five years later in 1919. The couple are buried together in St Machar's churchyard in Aberdeen.

An equally happy, but slightly odder pairing could be found in Samuel and Florence Baker. At thirty-eight, widower Sam Baker embodied the spirit of Empire and Englishness – a fine physical specimen of patriotism and plucky determination; a do-er rather than a thinker. Florence was a Transylvanian girl of seventeen, whom Sam purchased in a slave auction. When their two lives became entwined in Vidin (now Bulgaria) in 1859, Sam was on a hunting expedition with Maharaja Duleep Singh, whom he had befriended on a shooting estate in Scotland. Florence had been orphaned when Romanian marauders murdered her parents in the Hungarian Revolution of 1848 and was destined for the harem of the Ottoman Pasha of Vidin. To our twenty-first century eyes, their unusual pairing has all the hallmarks of a deeply troubling relationship. She was twenty years his junior, vulnerable, and barely into adulthood. He was a frustrated widower with four daughters who were only marginally younger than Florence – all of whom he had left in England under the care of his unmarried sister Mary, so that he could gambol about Europe with his chums. Good old Mary.

Yet, the relationship turned out to be transformative. Not just for Florence, whose future Sam had saved, but for Sam who would go on to make several early in-roads in the suppression of the slave trade in the equatorial regions of the river Nile. They gave each other a renewed sense of purpose; and although both were lost and unrooted; in each other they found a companionship that would last a lifetime.

Florence became Sam's lover, and when he announced he wanted to find the source of the Nile, she insisted she accompany him too. Whether she would question this decision when she was later in the throes of malaria-induced hallucinations is anyone's guess, but she was devoted to him, and he to her. For propriety's sake she was described as his wife and 'traveling companion' and her usefulness seems to have been a much-coveted prize amongst the exploring fraternity. One contemporary of Baker's wrote that Florence,

> proved to be of much service to him [Sam], assisting him in keeping his scientific records, and filling many other offices, such as most African explorers can only dream of...Clear- headed, courageous, and of decision that quickly manifested itself in action when emergency required, few women could have performed the duties which she took upon herself, and which she did so well.[69]

In 1862 they began their journey, taking no less than twenty-one donkeys, four camels, four horses and ten tons of grain to make sure that the expedition party (which also included forty-five porters) had enough to eat. There was much haranguing of the African beasts along the way (old habits clearly died hard) and a slightly awkward meeting with King Kamrasi of Uganda who told Sam if he left Florence at his court, he would provide Sam with a prettier wife (a suggestion that was met with a rather frosty reception from Sam, and 'a countenance almost as amiable as the head of Medusa,'[70] from Florence), but the couple waded through swamps, jungle and forests together, weathering innumerable hardships along the way. Florence's loyalty to her husband never wavered, and she fulfilled her wifely role admirably – becoming lover, friend, assistant, hostess and carer - all the qualities Sam needed to fulfil his own life ambitions.

Florence felt a strong sense of obligation, and for the rest of her life she devoted herself to Baker uncomplainingly. She was grateful, and she showed her gratitude by giving herself, heart, body and soul, to her husband, but it would be wrong to think of women like Florence as meek and obedient. They too were ambitious to share with their husband whatever life had to offer. It was their hidden strength that kept them going.

Sam did eventually make an honest woman of Florence (or Flooey, as he charmingly called her). She had more than proved her worth - and a lot more besides – and the couple were married on 4 November 1865 in London. Their final adventure together saw them returning to Africa at the request of Isma'il Pasha, the Turkish Viceroy of Egypt, on a military expedition to supress the slave trade around Gondokoro, an island trading station on the White Nile in Southern Sudan and prepare it for a more civilised mode of administration. This time, Sam would return as Sir Samuel Baker, Governor General of the Equatorial Nile and major-general of the Ottoman army. Florence, as ever, was by his side, serving as a medic to the troops. According to her entry in the *Oxford Dictionary of National Biography*, when the army was defeated at Bunyoro in Western Uganda, Florence was there, carrying rifles, administering brandy, holding umbrellas (the entry doesn't make clear exactly what she was planning on doing with those) and getting handy with a pistol.[71] Good old Flo, always dependable in a crisis.

The couple eventually stopped being intrepid and in 1873 they settled down to a quiet life at their home in Newton Abbot, Devon. By this time, Lady Baker, as she was now titled, had been fully accepted into fashionable society. The same year, Sir Samuel addressed the Royal Geographical Society with Florence, who was seated next to the Prince of Wales, the future Edward VII, and with 'whom she entered into a lively conversation.'[72] From her tragic early life as an orphan and near-miss life of slavery, Florence had burst through conventional Victorian boundaries in spectacular style – although her background prevented her from being accepted at court by Queen Victoria. Knowing something of their characters, one suspects neither Lady Florence nor her husband were overly upset by this slight.

Sam continued to travel over the next twenty years, mostly on jaunts to hunt big game, until his death of a heart attack in late December 1893. Florence followed him into their shared vault twenty-two years later in March 1916. Her obituary expresses 'profound regret in the town [Newton Abbot], where she was beloved by all who knew her. The 'Lady Explorer,' as she is referred to, was 'As intrepid as her distinguished husband…Her generosity was unlimited, although she was most unostentatious in all she did…her death will be a heavy and almost irreparable loss.'[73]

A contemporary interpretation of Florence's narrative is that she merely exchanged one type of captivity for another; the role of slave depressingly like that of wife at times; but they seem to have had a mutually beneficial relationship. Quite simply, they needed each other; Samuel gave Florence a purpose, and Florence enabled Sam to have a purpose. Her role - to stay by his side irrespective of the danger to her own physical well-being - defined her entire life. And it was a role she was happy to be recruited into.

But for every wife who was glad to follow her husband, there were several who were not so keen. They did so, regardless – after all, many of them were expected to emulate that greatest and most dutiful of wives, Queen Victoria - but the consequences could be disastrous for all involved.

Mary Moffat's story is particularly woeful – a cautionary tale of how one man's patriarchal and single-minded approach to attaining his goal brought misery to those closest to him. Mary Moffat was already well-accustomed to life in a remote region. Born in 1821, the daughter of Scottish missionary Robert Moffat, she spent her early life at a mission station in Kuruman in the Northern Cape. It was here that she eventually met and married medical missionary, and soon-to-be famous explorer, David Livingstone in 1845. As far as David was concerned, Mary was a catch – she had been brought up in the bosom of evangelicalism; she was well educated and an excellent linguist; and she could bake bread and manage a household. She was everything a missionary husband could want. David was thrilled, writing to a friend that his new wife was, 'a matter-of-fact lady, a little thick, black-haired girl, sturdy and all I want.' Coming from Livingstone, this was praise indeed, despite the unfortunate turn of phrase. Writing to his mother after the wedding, Livingstone reveals that, 'Only yesterday I said to my wife, when I thought of the nice clean bed I enjoy now, 'You put me in mind of my mother; she was always particular about our beds and linen.'' Mary's response is not recorded.

Mary would need a certain amount of sturdiness to get through the trial of being married to David Livingstone. David was passionate about one thing, and one thing only – converting the African people to Christianity. It was a never-ending quest that would see the couple and their ever-expanding brood of children come uncomfortably close to prematurely meeting their maker on several occasions. Mary was destined to marry a missionary, having spent her life up to that point immersed in her faith and living her own nomadic life at the remote missionary station run by her Scottish Congregationalist parents. History would erase Mary's contribution to the Livingstone legend, yet it was Mary's maiden name and status as the daughter of the well-respected and influential Robert Moffat - as well as her ability to converse fluently with the native people - that would ease Livingstone's passage across Africa.

Following a brief period of domesticity in their first marital home - the mission station at Mabotsa - Livingstone was keen to start spreading the word of the Lord, and within months of the birth of their first child in 1846, the Livingstones were on the move, trekking through Bechuanaland to Chonuane, forty miles north of Mabotsa. All three were in less than perfect health - Mary's condition exacerbated by another pregnancy – and when their crops failed due to drought, they were forced back to Mary's family home in Kuruman where Mary, remarkably, given her weakened state, gave birth to a healthy daughter.

God wouldn't wait though and before long, the family of four were embarking on another missionary quest to Kolobeng, on the edge of the Kalahari Desert. They arrived in 1848, and Mary busied herself setting up home for herself and her two children in a crofter's cottage. In some respects, it was the closest thing Mary ever got to a permanent abode, but she would spend little time in it with her husband because before long he was striking out again – this time to find the divine vision of a 'remote and shining lake' about which he had heard. This would mean crossing the Kalahari. Luckily for Mary (on this occasion at least) common sense prevailed and she and the

Livingstone children, who now numbered three, were packed off to the in-laws back in Kuruman.

David Livingstone was away for six months, which no doubt gave Mary some respite and the children some semblance of normality in their anything-but-normal lives. However, when David returned, he was filled with enthusiasm for a new idea. He had found the lake (Ngami, on the Okavango swamp) and believed he could convert the Makololo people living in that region. But... he needed Mary and the children to go with him because it would smooth his evangelical path with the locals if he arrived as a 'family man' rather than a 'lone zealot.' Mary didn't hesitate to do his bidding, even though it meant a 1500-mile trek across the Kalahari with three small children. And also giving birth in the middle of nowhere - because she was pregnant again.

The Livingstones made it to Lake Ngami but two of the children were soon stricken with malaria and they were forced to retreat to their old home at Kolobeng. It was here that Mary gave birth to another daughter, but the baby survived for only six weeks and was buried in the dusty landscape, the first Christian burial in that region. Livingstone was not without heart. 'Our last child, a sweet little girl with blue eyes, was taken from us to join the company of the redeemed,' he wrote to his parents in 1850. 'It is wonderful how soon the affections twine round a little stranger. We felt her loss keenly.' Mary herself was dealing with the double blow of both grief for her lost child and the lingering effects of pregnancy and labour in such challenging circumstances. The birth of her daughter had caused her to suffer from a partial paralysis of her face. Yet Mary's commitment to her faith was undiminished – during the periods that she was settled in Kolobeng she ran the local mission school and was a well-known and well-liked figure in the community.

Undeterred by all that had already passed, and with their faith still intact despite the setbacks, the family set off across the Kalahari again in April 1851. 'I must wend my weary way into the far Interior, perhaps to be confined in the field,' Mary wrote to her mother.

Mrs Moffat was outraged and sent a strongly worded letter to her son-in-law. 'Was it not enough that you lost one lovely babe, and scarcely saved the others, while the mother came home threatened with paralysis?' she asked, angrily. 'And will you again expose her and them in those sickly regions on an *exploring* expedition? All the world will condemn the *cruelty* of the thing to say nothing of the indecorousness of it. A pregnant woman with three little children, trailing about with a company of the other sex, through the wilds of Africa, among savage men and beasts!'[74] Mrs Moffat had the measure of the situation, and that she – an experienced missionary traveller and dutiful wife herself – would balk at his actions says something of the gravity of the situation.

But it was too late. The Livingstones were already on their way across the bleak, arid deserts. It was here, in the wilderness, that the Livingstones were at their lowest ebb. The journey took them around the bed of the River Zouga and across an area of desert that was eerily devoid of life – so much so that 'not a bird or insect enlivened the landscape.'[75] It was so dismal that even Livingstone later remarked that, 'It was, without exception, the most uninviting prospect I ever beheld.'[76] Their guide, Shobo, accidentally took them in the wrong direction and then vanished completely, then they spent four days without water. Now painfully aware of the danger he had placed his family in, Livingstone records that,

> The supply of water in the wagons had been wasted by one of our servants, and by the afternoon only a small portion remained for the children. This was a bitterly anxious night; and next morning, the less there was of water, the more thirsty the little rogues became. The idea of their perishing before our eyes was terrible; it would almost have been a relief to me to have been reproached with being the entire cause of the catastrophe, but not one syllable of upbraiding was uttered by their mother, though the tearful eye told the agony within.[77]

How Mary managed to stay silent in the face of such hardship is, frankly, remarkable. That she managed to stay silent regarding her husband's decision to drag them across such an inhospitable and dangerous landscape is even more remarkable.

Mary's worst fears were realised in September 1851, when she was indeed 'confined in the field,' and gave birth to her fifth child – a son – allegedly under a camel thorn bush although Livingstone's own journals merely record the child was 'born at a place we always call Bellevue.'[78] Three days later, one of their sons, Thomas, was stricken with a fever which raged on and off for the following two weeks. Besides this small entry, Livingstone's journals dedicate the following three pages to observations of the weather, the landscape, the constitutions of the locals, the lack of game, correspondence from the Directors of the missionary society, and criticism of fellow Scotsman Roualeyn George Gordon Cumming's latest book ('a miserably poor thing'[79]). Livingstone's lack of concern for his family may seem heartless – callous, even. Yet it would be disingenuous to assume Livingstone had no heart. Quite simply, his faith gave him the reassurance he needed to press on regardless, sustained by the belief that 'If God has accepted my service, then my life is charmed till my work is done.'[80] Like Annie Taylor, whose unswerving faith meant she would always be 'quite safe here with Jesus,' Livingstone believed he and his family were protected by God. In his mind, he was God's servant and God would not forsake them. That was the theory, anyway. In reality they were all mortal, and as such were subject to the same fate as everyone else – heathen or Christian.

On their return journey in 1852, Livingstone had time to consider his options, which led to a belated epiphanic moment – perhaps it would be better to leave his family somewhere safe after all, whilst he explored the untapped areas of the Zambezi. Mary was to travel back to Britain, with the children. This led to one of the oddest displays of behaviour to date. Livingstone escorted the woeful little group to the Cape and, putting them on a ship that was bound for England, wrote in a fit of uncharacteristic frenzy in his journal that 'to orphanize my children will be like tearing out my bowels, for they will forget me.'[81] When one of the children asked when they would see their home in Kolobeng again, Livingstone recorded that he had cried out, 'Never! The mark of Cain is on your foreheads: your father is a missionary!'[82] which seems a little histrionic, not to mention unfair on the poor child.

Mary was to have four years respite in all, but despite improvements in her health and that of her children, she was restless to return to her husband's side. Livingstone returned to Britain for two years in 1856; his star very much on the ascent whilst Mary's had waned to the point of invisibility. David was received by Queen Victoria and publicly lauded for his contribution to exploration but had further ambitions – to discover a trade route up the Zambezi and halt the slave trade. This time, Mary was to accompany him, along with their youngest son, six-year-old William Oswell. But within a month of them leaving British shores, Mary was pregnant again. 'This is a great trial to me, for, had she come with us, she might have proved of essential service to the Expedition in case of sickness and otherwise, but it may all turn out for the best,'[83] Livingstone recorded in his journal, without an ounce of guilt. Mary was once again shipped off to her parent's home in Kuruman, no doubt much to the relief of Mrs Moffat.

She then returned to England for a time before her final, fated return to Africa in 1861. This time she travelled without her children, in the company of several new members of the Mission – many of whom succumbed to malaria within months of reaching the Zambezi. Mary was reunited with Livingstone on 1 February but within weeks she had fallen ill. Mary slipped into a coma in Shupanga, on the banks of the Zambezi, and died on 27 April 1862. 'The man who had faced so many deaths, and braved so many dangers, was now utterly broken down and weeping like a child,'[84] wrote one of the doctors who was present during Mary's final hours. In a state of bewilderment that the inevitable has finally happened, Livingstone finally gives a rare glimpse of his depth of feeling towards his long-suffering wife. 'It is the first heavy stroke I have suffered, and quite takes away my strength,' he writes. 'I wept over her who well deserved many tears. I loved her when I married her, and the longer I lived with her I loved her the more. God pity the poor children, who were all tenderly attached to her, and I am left alone in the world by one whom I felt to be a part of myself.' Mary submitted herself to her husband's will and paid the ultimate price. As for Livingstone, he seemed to take for granted that Mary would always be by his side ('I have felt her to be so much a part of myself that I felt less anxiety for her…she seemed so strong too,' he wrote to her parents) - until, of course, she wasn't. Livingstone felt Mary's death keenly, though. His letters to his family announcing Mary's death speak painfully of her final hours – particularly one to his father-in-law, to which he adds a postscript the following day describing his grief: 'I feel greatly distressed at times and weep bitter tears…'[85]

The climate and logistics of transporting a body across Africa necessitated that Mary be buried where she died. 'We have not the proper chemicals to make it back - I placed a cross over her tomb a sacred symbol in these regions and have sent for a gravestone with the inscription on one side English and on the other Portuguese 'Here repose the mortal remains of Mary Moffat Livingstone in hope of a joyful resurrection by our Saviour Jesus Christ – died 27th April 1862 aged 41 years.' She rests in the Mozambique village of Chupanga, her grave signposted from the road as *Tumulo da esposa Dr David Livingstone* - the grave of the wife of Dr Livingstone. Even in death, Mary plays the understudy.

But for every Mary Livingstone and Florence Baker, there were a handful of spirited ladies whose foremost duty was to themselves. If they were looking for excitement – and a little mischief too – their travels gave them the perfect opportunity for autonomy, often for the first time in their life.

Part 6

For the sheer fun of it

> This book is not to be mistaken for an apology. Not at all…. It is the recollections of a woman who is no longer young and who had crowded a great deal of movement and fun and action and love and adventure into a lifetime now drawing towards its close.
>
> And if I could live it again, this very long life of mine, I would love to do so. And the only difference would be that I would try to crowd in still more…more places, more things, more women, more men, more love, more excitement. Let the Mrs Grundys arch their eyebrows and reach for their smelling salts.[1]
>
> Aimée Crocker, 1936

If any nineteenth-century lady was most deserving of some excitement and *joie de vivre* in their life, it was Austrian explorer Ida Pfeiffer. Ida would rack up an astonishing 32,000 kilometres by land and 240,000 kilometres by sea on her travels through Southeast Asia, the Americas, the Middle East, and Africa. She also managed to fit in two trips around the world between 1846 and 1855, when most ladies of her age were content to sit by the fireside darning socks. Her difficult, and at times eccentric childhood gave her all the impetus she needed to go out and enjoy the world, which is exactly what she did at the first opportunity.

Born in 1797 in a family that was already grappling with five healthy boys, Ida spent her childhood dressed as a boy and enjoying all the same wholesome male activities that were available to her brothers. She was educated alongside them and joined in with their games and pursuits. 'I was not shy,' she wrote in her posthumously published autobiography, 'but wild as a boy, and bolder and more forward than my elder brothers.'[2] Ida decided early on that she wanted to travel and explore, writing later that, 'When I was but a little child, I had already a strong desire to see the world. Whenever I met a travelling-carriage, I would stop involuntarily, and gaze after it until it had disappeared; I used even to envy the postilion, for I thought he also must have accomplished the whole long journey.'[3]

Imagine Ida's dismay then, when on her father's death in 1806, her mother attempted to instil some feminine charms into the gender-confused nine-year-old by forcing her into petticoats and undoing the 'boyish' education she had so far received. When Ida fell ill as a result, she was given a four year stay of execution and was allowed back into her boy's clothes on medical grounds, but by age thirteen she was expected to conform to social norms and the petticoats were thrust back on her.

To escape the influence of her mother, she married Dr Mark Anton Pfeiffer, a lawyer twenty-four years her senior and a widower with a grown-up son. The couple had two children of their own in 1821 and 1824. Unfortunately for everyone, Dr Pfeiffer made the colossal mistake of winning a case against several corrupt senior government officials in Galicia (in Spain) and was forced to resign. His actions made it impossible for him to secure further employment and the couple were forced to live separately, with Ida moving back and forth between Vienna, where her children were, and Lemberg where her husband was living. To support her family, she gave drawing and music lessons, and borrowed money from her brothers but the family spent many years teetering on the edge of poverty. The situation eased slightly on the death of her mother due to a small inheritance - and when her children left home and her husband died in 1838, she was finally able to fulfil some of her dreams. It was 1842, and she was by now forty-five, but she was finally free. Even the thought of dying on-route wasn't enough to stop her and she made a will and settled all her affairs before embarking on her first journey to the Holy Land, setting off down the Danube on 22 March.

Her first foray into exploration was not without incident. Apart from the usual gripes regarding cleanliness of the streets ('the town promises much but proves to be just such a miserable dirty place'[4]), and the unappealing nature of the local coffee houses ('…if it were not for the people sitting in front of them drinking coffee and smoking tobacco, no one would do these dirty rooms the honour of taking them for places of entertainment'[5]), Ida was simultaneously appalled and amused by the attempts of the Galatz officials to control the spread of cholera. In a scene that wouldn't have looked out of place in the worst days of the Covid pandemic in 2020, Ida found herself negotiating the 'social distancing' rules of international travel circa 1842. 'A wooden railing forms the barrier between the healthy people and those who come from or intend travelling to a country infected with the plague,'[6] she records. 'Whoever passes this line of demarcation is not allowed to return. Soldiers, officers, government officials, and superintendents, the latter of whom are armed with sticks and pairs of tongs, stand at the entrance to drive those forcibly back who will not be content with fair words….A gentleman on the 'plague' side wished to give a letter to one on the other; it was immediately snatched from his hand and handed across by means of a pair of tongs… 'Pray hand me over my luggage,' cries one. 'Keep farther away; don't come near me, and mind you don't touch me!' anxiously exclaims another.'[7]

Catching cholera was just one of many perils to the nineteenth-century traveller, but even more life-threatening were the perilous sea journeys and cross-country treks they undertook to reach their destinations. A large part of being an explorer is a certain amount of acceptance that death is around every corner – if not from disease, then from some other accident or mishap caused by the arduous nature of the journey. Fatalism is in the explorer's genetic make-up, or at least in their 'devil may care' attitude to their endeavours. Ida clearly had this in spades, deciding that on the leg of her journey by boat over the Black Sea that she would defy nature by going up on deck mid-storm. 'Holding tightly on,' she writes, 'I bade defiance to the waves, which broke over the ship and wetted me all over, as though to cool my feverish heat. I could now form a clear and vivid conception of a storm at sea. I saw the waves rush foaming on, and the ship now diving into an abyss, and anon rising with the speed of lightning to the peak of the highest wave. It was a thrilling, fearful sight...'[8] It was also a character-building exercise of the most terrifying kind.

Still, she made it to Constantinople alive, where she was unimpressed by the centuries old ritual of the dancing dervishes. The music was not to her taste: 'I do not remember ever to have heard a performance so utterly horrible. The instruments were a child's drum, a shepherd's pipe, and a miserable fiddle. Several voices set up a squeaking and whining accompaniment, with an utter disregard of time and tune;'[9] and neither was their dance; 'Twelve dervishes now began their dance, – if indeed a turning round in a circle, while their full dresses spread round them like a large wheel, can be called by such a name.'[10] She had a similarly strong reaction to the 'howling dervishes' of Scutari who tested her feminine sensibilities to the limit with their 'horrible and revolting spectacle,'[11] to the point where she feels it necessary to include a health warning advising persons afflicted with weak nerves not to witness the ritual 'for he certainly could not endure the sight.'[12]

On her first trip, Ida crammed in an enormous number of excursions. Perhaps she was nervous she'd never get the opportunity again. She visited a slave market, admired funeral customs, visited mosques, contemplated dressing as a boy again but decided against it, visited barely habited islands that now welcome millions of tourists a year (such as Cyprus), encountered Bedouins whom she was surprised to find weren't at all suspicious (despite loitering around the local ruins), visited Jerusalem and Gethsemane - as well as what seems like every church in the Holy Land – and saw off some would-be attackers. Having thoroughly explored the entire region of the Holy Land and Egypt she returned to Austria in December 1842. 'During my journey I had seen much and endured many hardships; I had found very few things as I had imagined them to be,' she wrote in the conclusion to her travelogue. This desire to *know* the world would inspire Ida to plan her grandest adventure yet, a trip around the globe, but before that she would need to acquire the all-important funds for her next 'mini-break' – this time to Sweden, Norway and Iceland. Despite professing a total lack of literary skill, Ida managed to pull off a publishing coup in 1844 by releasing an anonymous account of her journey in two volumes called *Reise einer Wienerin in das Heilige Land* ('A Vienna woman's trip to the Holy Land'). In return she received 700 guilders, which would help to fund her next trip. The book was an instant success. It went through three editions and was translated into Czech in 1846 and English in 1852.

Ida was determined to embark on her Nordic adventure with prior knowledge of the natural environment she would encounter on her travels. She learned animal and plant specimen preservation, as well as techniques using daguerreotypes – an early process of photography that was still in its infancy. She also learnt English and Danish. With these skills accomplished, by April 1845 she was ready to go exploring again.

Curiously, the preface to Ida's subsequent travelogue, *A journey to Iceland and travels in Sweden and Norway* (1852) offers an apology for being a woman who wants to explore. Feeling the need to explain why she chooses to spend her time travelling, Ida writes, 'When I was a little girl of ten or twelve years old, no reading was so attractive to me as books of travels…I could not but repine at the happiness of every great navigator, or discoverer, who could explore the yet unrevealed secrets of the natural world.'[13] Ida then explains at length her earlier travelling experiences with her family, and the process at which she reached her decision to go exploring. She closes the preface by imploring the reader to put their assumptions about her to one side: '…dear reader, chide not if I have written so much about myself, but let these, my inborn sentiments, plead my excuse for that love of adventure, which in the eyes of many does not accord with what is becoming in my sex. Judge me not too harshly, but rather freely grant me an indulgence, which injures no one and makes me so truly blessed.'[14]

Knowing something of Ida's financial situation, it is unlikely her request for understanding was not without additional motivation. Her message is aimed at her primary audience - the female readers who would be purchasing her book. To be deemed to be 'acceptable' to this group was crucial for the future success of her travel writing.

And if she couldn't sell her writing, she wouldn't be able to travel and explore in other parts of the world. Some criticism had already been levied at Ida's decision to travel on the publication of her first travelogue, so this was also one of the only ways she could hope to put the record straight. She may not have embarked on her initial trip to the Holy Land with the intention of publishing her journey; indeed, the press were keen to point out that 'she had no intention whatever of publishing her tour. It was not the love of fame or of money that induced her to commence it. Her reason was simply that the desire to travel had possessed her, and she could not stay at home.'[15] But once she realised there would be a market for her writing, which would provide the much-needed funds to pursue her dreams on a more permanent basis, it was a natural solution.

Ida's journey took her to Prague, Leipzig, Hamburg and Kiel to Copenhagen, and then on to Iceland, where she rode to Reykjavík on horseback and toured the geothermal area of Krýsuvík with a local guide. On her excursions, Ida visited the renowned sulphur springs and geysers, grottos and volcanoes. She climbed Mount Hekla - a treacherous journey even with the modern trappings of safety equipment – over 'steep masses of lava, sharp and pointed, which covered the whole side of the mountain. I do not know how often I fell and cut my hands on the jagged points of the lava. It was a fearful journey.'[16]

The climb may have been perilous, but it afforded Ida a chance to view the desolate landscape laid out before her. 'The whole country appeared only burnt-out fire. Here lava was piled up in steep inaccessible mountains; there stony rivers, whose length and breadth seemed immeasurable, filled the once verdant fields. Everything was jumbled together, and yet the course of the last eruption could be distinctly traced. I stood there, in the centre of horrible precipices, caves, streams, valleys, and mountains, and scarcely comprehended how it was possible to penetrate far, and was overcome with terror at the thought which involuntarily obtruded itself the possibility of never finding my way again out of these terrible labyrinths.'[17]

Her description is bleak, but one senses a slight thrill at the riskiness of it all. Ida's words betray the fatalistic outlook that is inherent in many a past (and present) explorer – an outlook that conversely, makes them feel more alive. Despite the ruinous nature of the landscape, she was still able to marvel at the stark beauty of the countryside – a landscape that seemed to contradict itself at every turn:

> On one hand are beautiful fields spread with herbage of a velvet green; and on the other, hills of black and shining lava. The meadows are traversed by lava-streams and patches of sand…On the left of the valley near Salsun, and at the foot of a hill, is a pretty lake, on whose shores reposed a flock of sheep. Not far from thence is a fine hill perfectly solitary and severed from the rest, as if it were banished and discarded by its neighbors [sic]. The whole of this landscape is completely Icelandic, and so peculiar and striking that it will be impressed for ever upon my recollection.[18]

The main draw for Ida were the spectacular geysers - or 'caldrons' as she refers to them – and having been issued with instructions as to when and how to spot their imminent eruption, she bravely spent a night alone in a tent with only the Geiser for company,

> I sat either beneath my tent, or in front of it, listening with stretched attention for the signs I had been told to expect. Towards midnight - the hour for spirits - I heard a few dull sounds, like those of a distant cannon, and rushing from the tent, I waited for the subterranean rumblings and the trembling and splitting of the earth, which, according to the books I had read, were the forerunners of an eruption. I could hardly defend myself from a paroxysm of fear; it is no slight thing to be alone, at midnight, in such a scene.[19]

An engraving that accompanies the text shows Ida's tent pitched perilously close to the geyser – little wonder she felt some 'rumblings.' Ida returned to Vienna from her Nordic adventure in October 1845 and published her journal to great acclaim the following year, with an English translation released in Britain in 1852. 'Mme. Ida Pfeiffer is a great traveller – the female Humboldt,' announced *The Globe* on the British release of her journal. 'The account she has given of the scenes she has witnessed and the perils she has passed through is highly interesting. The translation now offered of this 'Visit to Iceland' will be a great acquisition to the traveller's library.'[20]

Like many of her fellow explorers, she couldn't sit still for long. The urge to travel was far too strong, and the success of her journals gave her further encouragement – not that she needed any – and by the following year her bags were packed again. Exploring a region was no longer enough to satisfy Ida's insatiable desire to travel, so this time she planned a trip around the globe.

Ida left Vienna in May 1846, travelling first to Rio de Janeiro in Brazil and on into Chile. Ida begins her journal with her usual apologies and self-effacing comments, professing that:

> On the one hand, I possess too little wit and humour to render my writings amusing; and, on the other, too little knowledge to judge rightly of what I have gone through. The only gift to which I can lay claim is that of narrating in a simple manner the different scenes in which I have played a part, and the different objects I have beheld; if I ever pronounce an opinion, I do so merely on my own personal experience.[21]

Despite her modesty, Ida is keen to offer virgin travellers several tips – drawn from her own long-standing experiences, such as packing additional food (since the ships portions are measly), bedding for the unfurnished berth (Ida recommends a 'mattress, bolster, and counterpane'[22] at the very least) and, for those who are travelling with children, 'a goat as well.'[23]

Ida's excursions in Brazil afforded her the opportunity to collect specimens from the rich vegetation and she organised her explorations with a view to catching insects and taking cuttings from the surrounding landscape. On one unfortunate occasion, her wholesome activities were interrupted by a man wielding a knife and lasso, and it quickly became clear that he intended to murder Ida and her companion Count Berchtold, and drag their bodies into the forest. A very one-sided struggle ensued since the only weapons Ida and the Count had to defend themselves were their parasols. Ida brandished her plant-cutting knife at the assailant, 'fully determined to sell my life as dearly as possible,'[24] but to no avail. Her parasol was quickly hacked to pieces, 'leaving only a piece of the handle in my hand,'[25] she recounts, and she sustained two injuries to her left arm. Ida herself managed to inflict an injury on the man, wounding him severely in the hand, and the Count (whose actions until this point seem wholly ineffectual) suffered a cut across his hand having managed to grab their attacker from behind. Luckily for Ida and the Count, the horrifying experience was short-lived thanks to the providence of two men passing-by on horseback. At the sound of hooves, the assailant ran off into the forest, where he was eventually captured, beaten and removed to a nearby house.

After being patched up with bandages, the two travelling companions re-joined the path through the woods and continued their journey although 'not, it is true, entirely devoid of fear, especially when we met one or more negroes, but without any further mishap, and with a continually increasing admiration of the beautiful scenery.'[26]

More drama was to come as Ida made the notoriously perilous trip around Cape Horn (a 'trifling passage,' as she refers to it) to reach her next destination, Valparaiso in Chile. 'On the 3rd of February,' Ida writes, 'we were fortunate enough to reach the southernmost point of America, so dreaded by all mariners. Bare, pointed mountains, one of which looks like a crater that has fallen in, form the extremity of the mighty mountain-chain, and a magnificent group of colossal black rocks (basalt?), of all shapes and sizes, are scattered at some distance in advance, and are separated only by a small arm of the sea.'[27] Cape Horn did not disappoint, and over the next two weeks the sails were broken and ripped to shreds on several occasions – and had to be replaced by the crew as they battled to keep the ship upright. For Ida, the biggest inconvenience, apart from the continual rolling of the ship was the fact she couldn't sit at a table to eat her meals. Instead, the passengers were 'obliged to squat down upon the ground, and hold our plates in our hands.'[28] Ida narrowly escaped being scalded after a steward stumbled with a coffee pot and threw the 'burning contents' of the pot over her.

Having made it to the entrance to the Straits of Magellan, they assumed their ordeal was over, but a storm rapidly whipped up and they lost another four sails to the 'tremendous waves which broke with such fury over the ship, that they tore up one of the planks of the deck, and let the water into the cargo of sugar. The deck itself was like a lake, and the portholes had to be opened in order to get rid of the water more quickly.'[29] In the melee, Ida still managed to document the food situation, which now consisted of merely 'bread and cheese and raw ham, which we with great difficulty conveyed to our mouth.'[30] Next, she turned her irritation to the captain of the ship, who decided he wanted to make a grand entrance at the port of Valparaiso, meaning the entire ship needed to be renovated and painted in oil-colours. 'Not content with creating a most horrible disturbance over our heads, the carpenter invaded even our cabins, filling all our things with sawdust and dirt, so that we poor passengers had not a dry or quiet place of refuge in the whole ship,'[31] Ida complained. She begrudgingly accepted this situation, blaming the captain's unreasonable behaviour on his autocratic approach to running his ship and 'despotic power.'

The ship finally sailed into the port of Valparaiso on 2 March, where Ida stayed for two weeks before boarding a ship again – this time bound for the Island of Tahiti. After sailing for a further thirty-nine days (mostly without mishap and only one major storm) they arrived at the Tahitian port of Papeiti, where Ida mostly despaired of the native locals and what she perceived to be their entire lack of morals. She was also hosted by Queen Pomaré of Tahiti and her entourage of lesser royalty and ladies of honour. Ida thoroughly explored the island landscape and took a trip across a lake in a hastily erected boat ('he [Ida's guide] tore off some plantain- branches, fastened them together with long, tough grass, laid a few leaves upon them, launched them in the water, and then told me to take possession of this apology for a boat'[32]), before finally leaving Tahiti on 17 May. 'I felt extremely reluctant to leave,' she wrote, 'and the only thing that tended at all to cheer my spirits was the thought of my speedy arrival in China, that most wonderful of all known countries.'

Ida had a clear idea of what to expect in China – which she would have gleaned from the Orientalist depictions of Asia that abounded in the mid nineteenth-century. To many Europeans, China, at best, represented something exotic – a land of mystery, folklore and ancient peoples with an equally ancient set of customs. At worst, they perceived it as backward, peculiar and populated by a biologically inferior race. Ida pictured, 'the mandarins with their high caps, and the ladies with their tiny feet,' both stereotypes that she was delighted to see fulfilled during her travels. 'A year before my arrival in China,' Ida documents, 'it would have seemed hardly credible to me that I should ever succeed in taking my place among the small number of Europeans who are acquainted with that remarkable country, not from books alone, but from actual observation; I never believed that I should really behold the Chinese, with their shaven heads, long tails, and small, ugly, narrow eyes, the exact counterparts of the representations of them which we have in Europe.'[33] Ida sees her role as a passive observer, and the native Chinese are there to be studied and objectified. Just as her male counterparts take a paternalistic view of 'the native,' so too does Ida – thus allowing her to raise her own status to a more superior level – that of the scholar, a role that would be unattainable for her in Europe. Ida, of course, is not alone in her opinions – her writings are representative of the attitudes of many Victorian women who wrote about and studied the people of East Asia. Their accounts were written in a century of colonialism – an era that still casts a long shadow, the repercussions of which we are only really beginning to comprehend today. It is within this context we must pick apart and attempt to understand their narratives.

For all her excitement, Ida spent little time in China, and was there for just five weeks before leaving mainland China for Hong Kong, onboard an extremely disagreeable English steamer which had no decent cutlery, and then travelling onwards to Singapore. In Singapore, Ida decided to brave the jungle – and its many potentially dangerous inhabitants – and her excitement at her daring is evident when she saw the jungle creatures for the first time: 'My previous satisfaction was greatly augmented on seeing several apes skipping about on the highest branches of the trees, while others were heard chattering in our immediate vicinity. This was the first time I had seen these animals in a state of perfect freedom, and I secretly felt very much delighted that the gentlemen with me did not succeed in shooting any of the mischievous little creatures.'[34] What Ida really wanted to see, however, was a tiger. In this she was unsuccessful, but she did find a twelve-foot snake, which was promptly shot and gifted to their Chinese hosts, who intimated that 'they would not touch it.' However, Ida later wrote that she was 'convinced that this was all pretence, for on returning some hours later from our hunting excursion and going into one of their huts, we found them all seated round a large dish in which were pieces of roast meat of the peculiar round shape of the serpent.' Nothing and nobody could pull the wool over Ida's eyes, and now wholly convinced that she was staring at a plate of roast snake, she decided to try some herself. 'I found the flesh particularly tender and delicate,' she wrote, 'even more tender than that of a chicken.'

Ida's next stop was Ceylon, where she visited a Buddhist temple, then onwards to Madras and the mouth of the Ganges, which she eventually reached on 4 November 1847. From there she travelled up to Calcutta, trying a palanquin on the way which - unlike May French Sheldon and her 'white elephant' – she found a far from pleasant mode of transport: 'I was overpowered by feelings of the most disagreeable kind the first time I used a palanquin, I could not help feeling how degrading it was to human beings to employ them as beasts of burden.'

Having marvelled at the louche indulgence of the European population of Calcutta (now Kolkata) and their multitude of staff ('In some European families I visited there were from sixty to seventy servants, and from fifteen to twenty horses'), she describes an altogether more sobering experience – the Hindu death-house and rituals associated with death. 'I was horrified,' Ida writes. 'I hurried away, and it was long before I could efface the impression made upon my mind by this hideous spectacle.'[35]

From Calcutta, she sailed on a steamer to Benares where she dined with the Rajah and was entertained by two dancing girls who, she relates, 'shrieked so miserably that I was in fear for my ears and nerves.' The Rajah's extravagant lifestyle does not pass without a mention and Ida reveals with a total lack of decorum that 'the prince has forty wives, about a thousand servants and soldiers, a hundred horses, fifty camels, and twenty elephants' – one of which Ida enjoyed a ride on as part of her visit, giving her 'great pleasure.'

By April 1848, Ida was on her way to Baghdad on the steamer 'Sir Charles Forbes.' But there were no private cabins available, so Ida found herself creating a makeshift bed on deck, under the captain's dinner table, which she had to vacate every time a meal was served. Despite this hardship, and the trauma of having to contend with sickness and fever, as well as the death of three passengers from smallpox, Ida finally arrived in the historic port city of Bassora (now Basra in Iraq). 'Bassora, one of the largest towns of Mesopotamia, has among its inhabitants only a single European,' Ida writes, giving some sense of just how vulnerable she was – not only as a European, but as a woman. This sentiment is compounded when she was accosted by a group of women and children when she went searching for insects in the local undergrowth: 'I found myself on a sudden surrounded by a swarm of women and children, so that I thought it advisable to hasten back again to the ship's people – not that anyone offered me any violence; but they crowded round me, handled my dress, wanted to put on my straw bonnet; and this familiarity was far from pleasant on account of their extreme dirtiness.'[36] She herself became the 'other' - a spectacle to be gawped at by the fascinated natives. It was clearly not an experience that Ida enjoyed, despite her white privilege permitting her to do exactly that during her travels abroad.

From Bassora she travelled to Baghdad, where she visited a harem and a public bath – both of which horrified Ida's nineteenth-century sensibilities and filled her with 'disgust and commiseration,'[37] before touring the ancient ruins of Babylon, which were much more to her liking. Next on the itinerary were Mosul and the ruins of Nineveh, but to get there, Ida had to travel three-hundred miles on a mule. Ida describes this experience as 'a journey full of difficulties and dangers, without any convenience, shelter, or protection. I travelled like the poorest Arab, and was obliged, like him, to be content to bear the most burning sun, with no food but bread and water, or, at the most, a handful of dates, or some cucumbers, and with the hot ground for a bed.'[38]

Ida pressed onwards to Persia and Tebris, and by August 1848, she was setting out for Nakhchivan bordering Armenia, and then heading for Tbilisi, the capital of Georgia. She then crossed the Black Sea into the Russian Empire, where there was much bemoaning about the weather, before heading once again to warmer climes in Greece and Corfu. On 31 October 1848, Ida finally arrived back in Vienna and finished her narrative much as she started it – with a statement highlighting her humble writing credentials and her wish for 'a charitable judgment upon my book, which in simple language describes what I have experienced, seen and felt, and makes no higher pretension than that of being sincere and trustworthy.'[39]

Ida must have enjoyed herself, for she quickly planned to do it all again. Her *Second Trip Around the World,* which she began in 1851, took her briefly to London, then South Africa, Asia and, in 1853, across the Pacific to North America during the peak of the California gold rush.

She spent eighteen months of her trip exploring the Sunda Islands, which gave her the opportunity to get off the tourist trail and confront the reality of exploring in the true sense of the word. Whilst in Borneo she visited a Dyak tribe, an experience she describes with the same sort of pragmatism she uses to describe the streets of London. 'On the same day,' Ida writes, 'I paid a visit to another tribe farther on, and found little difference from what I had observed among the first, except that I had the pleasure here of seeing a pair of handsome war trophies in two freshly-cut-off human heads.'[40] She then goes on to explain their appearance in detail – a passage perhaps not best read aloud in the drawing room – and after pausing to shudder, ruminates on whether Europeans are so very different from the 'despised savages' she was amongst. 'Is not every page of our history filled with horrid deeds of treachery and murder?'[41] she asks. 'I do not think we Europeans can venture to say much about the cruelty of these ignorant savages, who kill their enemies (as we do), but do not torture them, and, for what they do, may plead in excuse that they are without the light of religion or of intellectual culture. Can we with a very clear conscience preach to them upon the subject of mildness, mercy, and aversion to bloodshed?'[42] It was certainly food for thought – just not of a particularly palatable kind.

Ida's fame was now well established – so much so that the press regularly featured updates as to her whereabouts and how far she had travelled. A letter from a Boston correspondent who allegedly spent the evening with Ida gives an amusing insight into her experiences in California. "Of all the countries I have ever visited,' said Madame Pfeiffer, 'of all the vile, immoral places I have ever seen or heard of in savage or civilised lands, the gambling saloons in California are the worst. I went there in company with friends; the doors were open, everything inviting entrance. Splendour in every form, temptation most subtle and powerful, combined to lure the soul and body to destruction - splendid curtains, carpets, exquisitely painted pictures, whose subjects were so impure that I involuntarily placed my hands over my eyes - wine, liquors of all kinds free, and to be had for the asking, all combined to lure the poor mortal to sin and death."[43]

Ida travelled south into Central America, visiting Panama, Peru and Ecuador. She then boarded a steamer for New Orleans in May

1854 and toured the Great Lakes before heading home to Vienna at the end of July 1855. Her account, *A Lady's Second Journey Round the World* was completed in 1855 and published in Vienna, before being translated into English in the same year.

Ida's final journey to Madagascar was her most dramatic and sensational - not because of the cannibalistic habits of the locals, but because she unwittingly became tangled in a political plot. On her way to the island she stopped at Cape Town in South Africa, where she met the French civil engineer and slave trader Joseph-François Lambert. Ida was unaware that Lambert had joined forces with industrialist Jean Laborde in a plot to overthrow the Queen of Madagascar, Ranavalona I, and unwittingly found herself in the centre of a political maelstrom, and although she played no part in the conspiracy she was expelled from Madagascar in July 1857 after the queen discovered the attempted coup.

Whilst her book is rich in many details of all the sights, sounds - and very often smells – that Ida encountered, she had an almost unnatural fascination with describing how women of colour appeared and behaved. In almost every location, she cast a judgemental gaze on the female inhabitants. In the town of Küferi on the route from Baghdad to Mosul, the women 'increase their natural ugliness, by dyeing their hair and nails reddish brown with henna, and by tattooing their hands and arms.'[44] In the Brazilian interior the Indians were 'still more ugly than the negroes...a peculiar look of stupidity is spread over the whole face, and is more especially to be attributed to the way in which their mouths are always kept opened.'[45] To a twenty-first century reader, such words are jarring, but this attitude – however distasteful now – can perhaps be framed in terms of Ida's own awareness of her new found importance. No longer is she the meek, self-sacrificing Austrian homemaker with no real power or influence; judged by a patriarchal society. Abroad, she holds the power and the patriarchal gaze, and by diminishing the subject of her observations to almost animal-like creatures, she also renders them less threatening. It is also worth noting that the English translation of *A Woman's Journey Round The World* coincided with Sophia Poole's *The Englishwoman in Egypt* (1851), which puts Egyptian women and their customs at the forefront of the narrative. Just as Poole presents 'detached sketches'[46] and invites her readers to 'amuse yourself by trying if you can to put them together,'[47] Ida prefaces all her travelogues with a sort of disclaimer that she is a mere observer and recorder of what she sees. Her role is 'narrating in a simple manner the different scenes in which I have played a part, and the different objects I have beheld; if I ever pronounce an opinion, I do so merely on my own personal experience.'[48] The things that female explorers and travel writers saw on their journeys often perplexed them. With no point of reference for harems, Turkish baths, and other immodest and indecipherable locations where native women could be themselves, explorers such as Ida Pfeiffer and Sophia Poole could only export what they saw onto the page and allow the reader to draw their own conclusions. They are passive observers, rather than active participants in the cultures they 'immerse' themselves in. Both women walked the tightrope between living on the margin of white European society, due to their decision to go roaming abroad, but also attempting to stay within its boundaries and adhering to its codes of conduct. Because to alienate white society and a potential readership by lauding native customs would have meant career suicide, and pariahdom.

It was a tightrope that Ida navigated successfully. One of the translators who worked on *A Woman's Journey Round The World* – a Mrs Sinnett – commented that:

> We have often heard of late years of a certain probably fabulous creature, supposed to exist in the latitude of Berlin and elsewhere, and denominated an 'emancipated woman'; nothing can be less like Madame Pfeiffer, yet truly she has emancipated herself in earnest – not from the fashions of gowns and petticoats, but from indolence and vanity and fear; under whose bondage both the stronger and the weaker sex are liable to fall, and that too, without deviating for a moment from the simplicity and

housewifely sobriety of her sex, her age, and her position in life. There is no country in the world where such a writer is likely to meet with more cordial appreciation than in England.[49]

And the English were ready to receive Madame Pfeiffer – so long as the English translation was re-named *A Lady's Voyage Round the World* (much more befitting) and all the boring bits were left out (much more exciting): 'whenever she [Ida] threatens to be tedious, Mrs Sinnett speeds her on her way. She has *not* translated verbatim the original, but has adopted such retrenchments as were found expedient, where the ground was quite familiar, and some omissions where the matters dwelt on were well known to the English, however strange to the German reader; this is to translate with taste, as well as with fidelity, for no thought is *omitted,* or even word capable of conveying pleasure or information…'[50]

In essence, Ida's travelogue was made more suitable for an English audience. What was lost in translation is difficult to determine without a thorough re-translation of the entire German text, but the essence remains – Ida was there to enjoy herself and have fun. So long as she did it tastefully.

Her pleasure seeking didn't extend too far into the grittier places she visited though, and her immersion in other cultures stopped at spending any more time than was necessary in some of the more impoverished settings she encountered. Where this was the case, she attempted to convey through her choice of language the impression that each place had formed, whether that be 'agreeable and handsome,' or 'wretched and dirty' – the latter is used more frequently.

There were of course practical reasons that explain her decision not to 'rough it' with the locals unless she had no other option. As a woman travelling alone in the mid-nineteenth century, Ida was vulnerable – both financially and physically. Without the patronage and protection of the learned societies, such as the Royal Geographical Society, on which many male explorers funded their adventures, Ida's safety was dependent on being in contact with, or at least introduced to, other Europeans in the countries she was travelling through. And when she was alone in a native community, which at times she was – particularly when negotiating the deserts of the Middle East – she was reliant on their goodwill to assist and protect her, even on a basic level such as providing her with shelter from the oppressive heat. 'How I envied the missionaries and scientific men,' Ida lamented whilst being scorched under the desert sun, 'who undertake their laborious journeys furnished with horses, tents, provisions, and servants.'[51]

Any further contact with the locals would be deemed 'improper' and would alienate her middle-class female readership – the very people Ida relied upon to buy her travelogues and fund her adventures. Practically, this meant navigating the expectations of a patriarchal world – a world where she needed to be perceived as behaving in a lady-like fashion and spending time in the company of those either of her own class or above it, but also allowing herself the freedom to just enjoy herself. This is never made explicit, of course. Ida is simply 'narrating in a simple manner' what she saw – and if she happened to have fun along the way, then that was simply a by-product of her travels. 'Fun,' however, would make her notorious, and that was not the image Ida was looking to project. Hester Stanhope, although dead for almost fifteen years, was still held up as a paradigm of unvirtuousness and posthumously capable of producing a shudder in even the most liberal of bosoms. Like all nineteenth-century female explorers, Ida had to be careful how she packaged up her motives, particularly given her sense of adventure. As one early twentieth-century study of Pfeiffer remarks: 'if a spirit like hers, so daring, so persevering, so tenacious, had been given to a man, history would have counted a Magellan or a Captain Cook the more.'[52]

So as far as motives went, flagrant adventuring was out. She also wasn't travelling for her health (Isabella Bird), or to preach the word of God (Mary Slessor, Annie Taylor), following a husband (Mary Livingstone, Florence Baker and Annie Boyle Hore), studying nature or antiquities (Marianne North and Mary Brodrick) or setting up a community (Catharine Traill). And neither was she attempting to seek fame or fortune a la Annie Londonderry or Nellie Bly. So, what exactly was she doing? 'Pursuing knowledge,' was Ida's official line – the world wasn't quite ready for ladies to have fun, but it wouldn't be long until women were feeling brazen and fearless enough to call it what it was; 'living adventurously;'[53] and those that did were far from bothered about justifying their motives in the first place.

* * * * *

As the long nineteenth century rumbled on towards its conclusion, a new phenomenon was emerging. It was time for the 'Old Maid' of the first half of the century to make way for the 'New Woman', who arrived just in time to enjoy the final years of the Victorian age, before having to mourn the loss of the of the era's greatest matriarch and its namesake – Queen Victoria.

The New Woman was sometimes (unfairly) characterised as a twist on the maiden aunt figure – she wore sensible clothes, rode a bicycle, was liberally minded, well-educated, probably smoked and could express her opinions on an equal footing to any man – particularly if they concerned the restrictions limiting the female sphere of influence to the home.

She was satirised by the same press who had bestowed on her her title, often appearing in illustrations next to her beloved 'safety bicycle' whilst a nearby gentleman makes a witty remark about her choice of clothing. For *Punch* magazine the New Woman was a figure of fun; an embittered, over-educated spinster who would be stuck on the shelf *ad infinitum*. "The Old Maid trying to be the Man," was just one description levied by a reader of *The Gentlewoman* amongst other equally unflattering terms, such as "A creature of opinions decided and skirts divided;' 'One who has ceased to be lady and has not attained to be gentleman;' 'The unsexed section of the sex;' 'Man's newest and best reason for remaining single;' and 'Madam become Adam.'"[54] These stereotypical and unattractive descriptions neatly obscured the truth – that the New Woman was young, vivacious, stylish, avant-garde, perhaps even a little shocking. Not only that – she was a threat to patriarchal stability, hence the mocking tone adopted by the male-oriented press.

Apart from providing fodder to fill the column inches of newspapers, the New Woman was represented in her many guises in art and literature. To English feminist writer Sarah Grand, one of the original exponents of the phrase 'New Woman,' she was a wife trapped in a loveless marriage, questioning the double standards of her husband's sexual past. To writer Thomas Hardy, she was a highly educated but socially isolated young lady, forced to marry her intellectual equal, rather than her true love; and for George Egerton (the pen-name of writer Mary Chavelita Dunne Bright) she was sexually adventurous and free to love whomever she wanted. She was often affluent, but not always, and she was just as likely to grace a factory floor as a drawing room.

It was against the backdrop of the Victorian fin de siècle that the most fun was to be had – particularly if you identified as a lady explorer. And even if you didn't, there was still much to be gained from seizing the day and going off on a little adventure. Because, why not?

Aimée Crocker aka Princess Palaikalani (Bliss of Heaven) of the Sandwich Islands had a string of real-life adventures that wouldn't have sounded out of place in one of Annie Londonderry's lectures. Born in 1864 in Sacramento, California, Aimée inherited $10 million (around $246,758,000 today) at the age of ten – a colossal sum of money even by today's standards. Her life story, which she details in her biography *And I'd Do It Again* (1936) is a eulogy for the louche and uninhibited. Amongst other 'accomplishments,' Aimée was famous for having a brief love affair with a Spanish bullfighter; surviving the trauma of a train crash, which claimed the lives of twenty-one people, escaping head-hunters in Borneo with a terrifying jungle river escape; being abducted by a Dyak prince; a lesbian double suicide; a poisoning in Hong Kong; a murder attempt by knife-throwing servants in Shanghai; and some bizarre sexual experiences involving a boa constrictor and a Chinese violin. She also established a Buddhist colony in Manhattan and managed to enrage a group of missionaries during her sojourn in the Sandwich Islands. During her travels, she also collected husbands (six were legal but she put the number at twelve), lovers, adopted children and snakes.

Naturally, all this gallivanting brought her to the attention of the press...and everyone else. 'Amy [sic] Crocker, today Mrs Harry Gilsey, owns three $10,000 dogs,' wrote the *San Francisco Examiner* in December 1899, 'and spends as much upon their food and blankets, medical attendance, and luxuries as would support several families. They are bathed in perfumed water in their private bathroom at the Gilsey House by their own dainty, skilful French maid...And they are having their miniatures painted by the famous animal artist, Mrs Lora C. Chandler, at $100 for each head.'[55] Dogs and husbands were a running theme for most of Aimée's newspaper coverage and if it wasn't Aimée's love life or obscene wealth being plastered all over the press, then her exploits abroad were always good fodder – and capable of raising more than a few eyebrows to boot.

Described as 'a petite blonde, of beautiful figure and rather pretty face, and a wealth of golden red hair,' Aimée was unapologetically adventurous and 'un-moral' (in her own words) and attributed her desire to experience all that the world – but most particularly Asia – had to offer to a vision she had as a child. 'The moonlight was pouring in the large bay window and shone directly on my bed,' Aimée described in her memoir.

> I could see very plainly, stretched out on my white bedspread, a woman. I knew she was very beautiful but I could not see her features because they were covered over with a veil of some gossamer material up to her eyes, and she was dressed in colored silk robes, a costume such as I had never seen, even in pictures. There she lay, radiant and shining her arms stretched back on the bed,

and looking straight at me. She smiled, I remember. She seemed to know me.[57]

Twenty years later, Aimée's vision led to a moment of realisation – that the vision was the calling card of Asia, and it was calling her to explore what she felt was her spiritual home. 'Very young indeed was I when the finger of the East reached out across the Pacific and touched me,' Aimée recalled. So, it was to the East she went, but first she would travel to Europe to be 'finished' where she fell in love with a German prince, and a Spanish toreador called Miguel - 'a real live toreador complete with be-buttoned and be-ribboned and be-caped costume…something to take the breath away.' Miguel was also 'brutal, conceited, haughty, passionate, direct, childish and completely irresponsible,' and it was with some relief that Aimée's mother managed to whisk her off to England, where passions were dampened, and everyone stuck to the accepted rules of drawing room etiquette. And so concluded Aimée's European 'finishing.'

None of the etiquette she learned seemed to stick, and the passionless and draughty drawing rooms of London had clearly taught her nothing. On her return to Sacramento, she fell in love with Porter Ashe, the son of a prominent San Francisco family, and they eloped and got married. Porter was twenty-one, Aimée just seventeen. The inevitable scandal that followed was only eclipsed by the drama of their honeymoon. Whilst crossing the Tahachapee Pass (now Tehachapi, California) on a Southern Pacific sleeper train from San Francisco to Los Angeles, the young lovers narrowly escaped death when the train that was carrying them left the track and plunged several hundred feet down a mountain pass. 'I came to,' Aimée recalled.

> It was black night, but everything about me seemed to be glaring red and yellow light. Fire. I moved my hand and felt the broken splintered end of what had been a steel beam. I knew the car was burning. I heard cries

and moaning about me. I dragged myself, dazed and frightened, to a sitting position, and then to my feet.[58]

Once reunited with Porter, the couple were to discover just how lucky they had been to survive the wreckage of the train, which 'now lay in flames, twisted and mangled like a huge broken snake.'[59]

> Bodies were being carried past us from further down. Wrecking crew, local farmers, unharmed men amongst the passengers, even the despised Chinese coolies, were doing their best to recover the dead and injured from the burning wreck. We were led back, up a difficult pass, to the railroad track where a flat car pushed by a dummy engine had rolled up. Shelter. The flat car was loaded with bodies, being piled up like lumber. The place was echoing with the moans of the injured and dying. It was insane, unbelievable, indescribable.[60]

What permanent effect this had on Aimée's mindset can only be guessed at. The press were quick to sensationalise the tragedy, even in the British regional newspapers, who reported victims being burned to death 'before the eyes of the survivors,'[61] or in one particularly gruesome account, 'literally roasted.'[62] Reports of the number killed ranged from seventeen, to twenty-one, with many never accounted for. Aimée herself put the number at 'forty-five killed and several hundred injured,'[63] but it's possible that Aimée's later reflections on the experience were unintentionally mis-remembered.

The marriage had started badly, and it was to end badly too. After travelling to Europe as part of their honeymoon, Aimée returned alone and gave birth to her first child – a daughter named Alma – in San Francisco, and her divorce from Porter was quick to follow.

Boredom and disillusionment set in. Fed up with 'civilised' life, Aimée decided to 'go away into the green, natural places, where men are men and women are all Hula-hula girls.' Put simply, she fancied a trip to the South Seas. 'It was the call of the East: I did not know it then, but I was listening attentively to that spell. My plan was to be alone, to do everything alone, to be independent;'[64] which for a woman of Aimée's means was entirely feasible. She was a divorcee; she was wealthy, and she had little else to do. After securing passage on the aptly named schooner *Tropic Star* – on the basis that she wasn't a missionary (the crew had had enough of them) she landed in Honolulu. She did not, as reported in the *Belfast Telegraph*, hide as a stowaway on a missionary ship – although that made a better story. Or live in a straw hut, which was also reported by the press at the time.

On arrival, Aimée was immediately struck by the island's natural beauty and purity, and the fact that King David Kalakaua didn't live in a 'ramshackle bamboo affair' but an 'immense modern stone building.'[65] Aimée's 'exploration' of the Sandwich Islands really only extended to visiting places where she could either swim or ride – two of her favourite pastimes – until she caught the attention of the local missionaries, who were thoroughly unimpressed by her frivolous use of time. 'The things they objected to were so silly,' she recalled. 'Just because I refused to live like an American…that I knew and liked many of the dancing-girls, dressed as they did when I wanted to, and learned to dance the hula-hula…and that I generally had a good time in my own and really harmless way, they were distressed and harassed and annoyed beyond measure.'[66]

Aimée was accused of being an 'immoral hussy' because she fraternised with the locals and was setting a bad example. According to the missionaries she was supposed to be acting superior to the locals, not cavorting alongside them. But Aimée was nothing if not forward-thinking. She recognised the double standards and called it what it was: hypocrisy. 'Are we not a little cocksure,' she wrote, 'a little conceited, in thinking that we only, who have specialized our lives to be in tune with automobiles, radios, cinemas… (and our drunkenness and organized crime and white-slavery, by the way) …are the only ones upon whom the Divine eye looks with favour?'[67]

Aimée was, however, quite happy to be 'given' an island by King David, complete with a local population of three-hundred people for her to rule. She was also bestowed with a Hawaiian name in honour of her new status – Princess Palaikalani aka 'Bliss of Heaven.' 'I did no actual ruling, of course, but I made frequent visits to 'my' island, and 'my' people were very sweet and amusing,'[68]

After a brief meandering around the Polynesian islands, and with that spiritual finger still beckoning her, Japan was the next stop. There, much flirting ensued – enabled by her new husband being called back to San Francisco on business – as well as much scandal. Aimée's love interest was a perfectly poised Japanese gentleman, Baron Takamini, who revealed the East to her 'as no one else could have done.'[69] She learnt a little about religion, and the Japanese attitude to sex and prostitution, but wastes minimal page-space on any more revelations, choosing instead to continue the story of the romantic dramas in her life. Against the exotic and heady background of Embassy balls, dinners and hobnobbing with the consulate, Aimée lost her Japanese lover to a fellow American, who broke Baron Takamini's heart by refusing to run away with him. Much to Aimée's horror, the lovelorn Baron committed suicide over Aimée's love rival. 'I lost a friend,' she writes. 'One of the best, one of the kindest, one of the most glorious men I have ever known.'[70]

Aimée was soon on the move again. After a quick pit stop to divorce her second husband ('our life together was not advisable… the divorce followed shortly. That is not an interesting story' Aimée divulges), she embarked for another Eastern location – this time Hong Kong. On the journey there she met a murderer who was in the process of poisoning her husband, a crime in which she succeeded and that Aimée, allegedly, only revealed thirty years later to the readers of her memoir. In Hong Kong, she experienced an opium den (one wonders what the missionaries would have made of that) and fell under the spell of a thoroughly unsuitable man called Huan Kai who, despite his 'culture and erudition,' she grew to be afraid of. This didn't stop her sailing with him in his yacht to Shanghai, a city 'like no other city in the world.' Despite several red flags, which Aimée herself recognised with the benefit of thirty years hindsight, she found herself trapped in a bizarre 'relationship' with Huan Kai, whose idea of a fun day out was to take her to the public execution of a local warlord.

> If I were a skilful writer, I might hope to create a shudder and make it tremble through these pages as I recall that day. My memory of it is still tinged with terror...not a fear of anything defined, but a formless fear for people, a realization of the savage in man which persists in spite of his speech, his race, his so-called culture, his customs or his law.[71]

More fun was had when Aimée visited a mysterious house where an old man was playing a violin and she experienced a sort of sexual awakening which she described later as though 'invisible hands were touching me...that sound...or was it music? ...enwrapped me and carried me out of myself into an orgy of physical hysteria.'[72]

Just when the reader thinks that things couldn't get much weirder, she reveals that Huan Kai tasked his guards to spy on her, and he began to insist they get married. Not a moment too late, Aimée recalled that two of his previous (wealthy) wives had died in mysterious circumstances and that he was also called 'Mr Gold' 'for no very pleasant reason.'[73] For Aimée, it was all starting to look a little disturbing. She managed to concoct an escape plan which involved hiding in a hotel, where she also dodged an assassination attempt on the same night. Her servant Meh-ki, who helped her to escape was not so fortunate as when she left the hotel room to fetch tea she didn't return. 'Her body was found at the end of the hall with a long knife through the back of her neck,' Aimée relates with aplomb.

The wealthy heiress then hastily left China lest she suffer the same fate, and travelled to Singapore via Hong Kong, a country she describes as 'a cauldron of boiling humanity,' before pushing onwards to Java and Borneo. But there was more trouble on the horizon. In Borneo she was abducted by a Dyak prince who wished to have her as a travelling companion – although Aimée herself conceded she was 'rather a willing prisoner' and saw the entire situation as a lark – and the pair travelled into the jungle. Another assassination attempt ensued (this time with poisoned arrows), meaning Aimée was forced to flee yet again before the Dyak 'head-hunters' caught up with her.

> It was night, and I could see nothing. I thought of crocodiles, animals that scream overhead and around the banks of the river, monkeys, and creatures that never existed in Borneo. I was in terror of everything, and especially of the lithe, powerful, bronzed, hairless men who were probably behind me in their swifter boats, fanatically pursuing me to carry my head triumphantly back for 'the good of the race.'[74]

Aimée drifted alone in a canoe for several hours with nothing but the jungle and river creatures for company. She finally landed in the Dutch outpost of Long Iram where she was offered temporary respite at the home of 'Captain van. B.' and after a short convalescence – made shorter still by the captain's desire to be rid of her – Aimée travelled onwards once again to India, the country whose 'living breath had been blowing upon me across broad seas, whose fingers had been beckoning me.'[75] It was there in Bombay, amongst the maelstrom of 'hundreds of other racial faces and costumes…like a colorful human cocktail,'[76] that she finally believed her quest to be at an end.

This was not the end of Aimée's adventures, however. They continued throughout her unconventional life, both in America and abroad – much to the delight of the press and an eager-to-read public. When she wasn't gallivanting around the East, Aimée populated her homes in the West with all the cultural icons of her spiritual home. Her Parisian 'Chinese Temple' home contained 250 Buddhas with emerald and ruby eyes lit by electric bulbs. 'It was reported that these eyes so played on the nerves of Prince Galitzin, the lady's fifth husband, that this was his reason for running away with her adopted daughter,'[77] reported *The Graphic* in 1927. Her most audacious stunt – and the one that was still talked about even after her death – was the reception she held for H. H. Kaa, Maharajah of Amber. All of New York society was invited to pay homage to her special guest – which turned out to be her twelve-foot pet boa constrictor, Kora. She was 'The Most Fantastic Woman of Her Age,' according to the *Philadelphia Enquirer*, whilst the *New York World* likened her to the Greek goddess Hebe, the daughter of Zeus and goddess of eternal youth. In the wake of her whirlwind life, she left a trail of broken hearts and unorthodox principles that paved the way for all the free-spirited Bohemians of the next generation.

It is hard not to like Aimée. Her memoir reads like a series of short fictional tales – much like her own book of short stories (or 'arabesques' as her publisher referred to them), *Moon-Madness and Other Fantasies* (1910). Aimée referred to the stories in *Moon Madness* as 'memoirs in fiction form,' which could just as easily describe her actual memoir *And I'd Do It Again*. There is certainly a great deal of overlap in the themes of both works – pearls, snakes, Maharajahs and rooms full of Buddhas. As well as a certain amount of dissatisfaction with men, which she explores in her story 'Betty and the Buddha,'

> Betty was tired of Bob, And Bert. And Reggie.
> So she took up with Buddha.
> Bob was too hot.
> Bert was too cold.
> And Reggie was neither hot nor cold – merely lukewarm.
> Men seemed all wrong.
> But Buddha seemed just right.
> One need never worry about him.
> He could always be relied upon.[78]

All the iconography associated with Aimée herself is evident in her writing. Whether her lived experience was exactly how she presents it in *And I'd Do It Again* is debateable. But irrespective of whether Aimée blurred the line between fiction and reality, one thing is certain – she was a restless spirit, whose explorations abroad afforded her the opportunity to explore herself, and the essence of what it means to be human. As she roamed the East, she also searched for a sense of who she was – and she found herself in the harems, the palaces of Indian princes, the opium dens, and the (many) scandalous adventures she professes to have had. Her vast wealth afforded her the freedom to act autonomously, and the choice to live unconventionally. As her biographer Kevin Taylor writes, 'Crocker's life is a cabinet of human curiosities, a celebration of some of the most eccentric, extravagant and extraordinary personalities…Her coterie was a pantheon of unique and fearless spirits.'[79] If Aimée collected anything on her travels, besides husbands and Buddhas, it was the ability to understand people – and through them, she began to understand herself. 'Husbands have little to do with people,'[80] she said famously, when challenged as to why her memoir virtually ignores her five marriages. Aimée married her fifth and final husband, the Russian Prince Mstislav Galitzine, when she was sixty-one and he was twenty-six – although that number excluded Aimée's alleged marriages abroad. 'The prince is my twelfth husband,' Aimée once said, 'if I include in my matrimonial list seven Oriental husbands, not registered under the laws of the Occident.'[81]

Aimée slipped into obscurity after her death in 1941 – an outcome she certainly would have been annoyed about, although she managed to posthumously hit the headlines just over a year later in spectacular style. 'Crash! Goes Aimée Gouraud's Shrine on the Burma Road' the headline of the *American Weekly* read on 10 May 1942. 'The oddest shrine in the world has been destroyed,'[82] the reporter writes:

> A Japanese bomb, aimed at Burma Road, went wide of its mark a few weeks ago and demolished it. The airmen

may have regretted the incident as a waste of good metal and explosive, but if they had known more about their accidental target they would probably have cheered up.

For that shrine was dedicated to no Buddhist saint. In fact it had no connection with anybody even remotely saintly. It was a memorial to an uninhibited American heiress named Aimée Crocker Gouraud.[83]

The shrine was suggested by the son of one of Aimée's many royal connections in Burma (now Myanmar) to recognise her friendship with his father and her legendary status among the Burmese people. 'Aimée thought that would be nice, too, so not long afterwards, Nicholas Roerich, a friend, was commissioned to do the job,' the article explains. The shrine included bells, an altar, and a panorama of paintings depicting Aimée's Burmese invasion. That it would then be 'accidentally' obliterated by the Japanese during their attempts to cleanse the East of its Western influence during World War Two was opportune.

It was also dramatic, and ironic – a suitable postscript to what was a thoroughly dramatic and ironic life.

Epilogue

The journey home

S. S. Malwa, February 25 – We sailed from Pinang in glorious sunshine at an early hour this afternoon, and have exchanged the sparkling calms of the Malacca Straits for the indolent roll of the Bay of Bengal. The steamer's head points north-west. In the far distance the hills of the Peninsula lie like mists upon a reddening sky. My tropic dream is fading and the 'Golden Chersonese' is already a memory.[1]

Isabella Bird Bishop, 1883.

May French Sheldon wasn't supposed to make it home alive. In fact, the multiple doctors who examined her during the final leg of her journey in 1891 were horrified that she was proceeding with her plan to go home when she was evidently in a 'dying condition.' 'I was utterly powerless,' May wrote in her travelogue, 'and had to lie just where placed, dying, it seemed, by degrees, my poor brain half delirious.'[2] Several days before her fever induced delirium, May had been unceremoniously dropped in the raging swell of a river – twice. The second time, which happened during the rescue attempt from her first fall, caused her to land awkwardly on jagged rocks. She knew immediately that her adventure was over, at least for the moment, and that her life depended on her reaching the coast and sailing back to her husband Eli Sheldon and their home in England. But May hadn't reckoned with catching dysentery. Her porters carried her day and night through the swampy countryside of Rufa – managing forty miles in one day – in a light hammock. She drifted in and out of consciousness whilst her porters battled to keep her alive by whatever simple means possible. 'I seemed to be surely dying from the sequences of the injury to my spine, starvation, and dysentery,' she writes miserably.

One thought alone kept her going: 'to live at least until I should once more see my husband and reach my home.'[3] Even when she fractured her skull from a freak gust of wind that flung her off her mattress and against the railings of the ship, she kept the faith.

'Does she live?' was her husband's first question when her boat docked in Naples.

The answer was that she did. 'Friends and the dear faithful Jacques [her pet dog] overwhelmed me with such a welcome. The little town, the loved home, was redolent with a greeting, and brilliant with flags, among which conspicuously floated my own American flag, which had acted as a talisman throughout my safari.'[4]

Death was no match for May, and the dark spectre of eternal rest wasn't going to deter her from making further explorations abroad over the following four years. She died in 1936 at the age of eighty-eight. 'Was it worthwhile?' May asks herself at the end of her narrative. After 'serious retrospection,' she concedes that it was.

Like many of her kind, May was at home in body, but her spirit was elsewhere. An explorer's homecoming is often a happy affair, but not for long, and the wanderlust that sent them travelling in the first place is never far from their thoughts. Ida Pfeiffer and Isabella Bird knew this better than anyone. This is never more evident than in those whose return to normality meant a return to domesticity and a slightly less glamorous existence. Mary Eyre, a poor and genteel mid-nineteenth-century traveller to the south of France, certainly had the shine taken off her wanderings abroad when she returned to Britain. 'London, by comparison, looked dark and dreary,' she writes. 'Its aspect depressed my soul – with its dingy rows of houses, it's densely populated streets, its crowds of pale-faced, slovenly looking men in shabby coats, its care worn, dirty women, in torn, draggled gowns, and faded bonnets, with dirty artificial flowers under them.'[5]

Little wonder they didn't hang around for long - the only remedy was to keep on moving, whether that meant returning to a country that had bewitched them or treading a new path somewhere else. For Amelia Edwards this meant the excitement of 'Untrodden Peaks and Unfrequented Valleys;' for Mary Kingsley, Africa would always be her mistress. 'The charm of West Africa is a painful one,' she wrote in 1897. 'It gives you pleasure to fall under its sway when you are out there, but when you are back here it gives you pain by calling you.'[6] In Mary's case, the pleasure-pain paradox ended with the latter. She died on the continent that she loved, whilst nursing prisoners-of-war in Simon's Town during the Second Boer War.

Those who managed to drag their weary, journey-fatigued limbs all the way home often felt a lack of belonging to the country they had left. 'I was now practically an exile from my own country,'[7] Margaret Fountaine lamented on her return to the home she owned in her later years in California, although she was grateful to the 'splendid land' for providing her with a home to go to. Their disconnected existence, roaming over land and sea and traversing the globe in all the directions of the compass left them disorientated and with no real affinity to any one place. 'I feel like a woman without a country,'[8] wrote Alexandra David-Neel, when she went to join her long-suffering husband in France. Settling down in one place also prompted reflection, and a dawning realization that an elderly failing body may not carry them much further – especially those who spent a lifetime travelling; only stopping when their health finally gave out and forced them into submission. 'It's a sad thing to grow old,' Alexandra David-Neel wrote at the age of seventy-five. 'I should have died in my tent in the Tibetan solitudes.'[9]

Some never gave up the hope that they may one day travel again. In some ways it sustained them that little bit longer. Isabella Bird kept her possessions in storage and was already packed for another trip to China in her final months – a journey she never got the opportunity to take. Being unable to fulfil their travelling ambitions, through physical or mental fatigue, was every explorer's greatest fear. In some respects, death was preferable – for wasn't it another journey? The ultimate chance to experience the unknown? Places unseen, native tongues unspoken, and paths not trodden; *these* were the explorers greatest concerns when their bodies started to fail them. Little wonder that they found solace in documenting their journeys for posterity.

And thank goodness they did because what they leave behind is a wealth of narratives that capture the lived experience of a rare breed of Victorian lady. Some of these remarkable ladies may well be forgotten in all but the smallest footnotes of history, but their legacy lives on in every woman who dares to believe she can do something different. They speak through every woman who has uttered the words, 'I can, and I will.' It is impossible to do all of them justice, of course. Whilst the ladies that dared to go where others wouldn't make up a very small coterie within the female population of the nineteenth-century, there are still too many to cover in any detail in only one tome. With so many larger-than-life characters, they jostle for attention and page space. In the case of this book, whoever shouted the loudest was included, but invariably many had to be left out due simply to a lack of space. Really, this book ought to be titled (*Some*) *Nineteenth Century Female Explorers* for it only features a small sample of the many narratives that have become known over the last fifty or so years – and no doubt there are many more, just waiting to be rediscovered and revisited.

In some respects, we ought not to be surprised by the pluckiness of their endeavours. The zeitgeist of the nineteenth-century was one of industry, aspiration, enterprising spirit, innovation and rapidly changing attitudes at all levels of society. Little wonder women wanted a slice of the action. Particularly later in the century, the emergence of education and career opportunities for women as well as new legal rights regarding property (although they would have to wait until the twentieth century to vote), meant more freedom and choice – and as a result, more opportunities to break down the barriers preventing them from acting independently. Did the fabled 'Angel in the House' ever really exist in human form? Perhaps, but I for one like to think that in every 'angel' lurked at least some grit and ambition to leave the domestic sphere and strike out on her own terms. After all, angels need to spread their wings every now and again, don't they?

Notes

Introduction

1. David Neel, A. (1927) *My Journey to Lhasa: The Personal Story of the only White Woman Who Succeeded in Entering the Forbidden City.* William Heinemann Ltd., London.
2. Ellis, Mrs S. (1839) *Women of England.* Fisher, Son & Co., London.
3. Ibid.
4. Ibid.
5. *Punch Magazine,* Vol. 104, Issue 2709, 10 June 1893.
6. Ibid.
7. Kingsley M. (1897) *Travels in West Africa.* Macmillan & Co., London.
8. Dixie, Lady Florence (1880) *Across Patagonia.* Richard Bentley & Son, London.
9. Lowe, E. (1857) *Unprotected Females in Norway; or the Pleasantest Way of Travelling There, Passing Through Denmark and Sweden with Scandinavian Sketches from Nature.* G. Routledge & Co., London.
10. Mazuchelli, N. E. (1876) *The Indian Alps and How We Crossed Them.* Longmans, Green, & Co., London.
11. Morris, M. with O'Connor, L. (Eds.) (2007) *The Illustrated Virago Book of Women Travellers.* Virago Press, London.
12. French Sheldon, M. (1892) *Sultan to Sultan: Adventures among the Masai and other tribes of East Africa.* Arena Publishing Co., Massachusetts.
13. Kabbani, R. (1988) (Eds.) *The Passionate Nomad: The Diary of Isabelle Eberhardt.* Beacon Press, Boston.
14. Ibid.

15. Meakin, A. (1901) *A Ribbon of Iron.* Archibald Constable & Co., London.
16. Campbell Davidson, L. (1889) *Hints to Lady Travellers: At Home and Abroad.* Illife & Son, London.
17. Duncan, J. E. (1906) *A Summer Ride Through Western Tibet.* Smith, Elder & Co., London.
18. Ibid.
19. Bird, I. L. (1883) *The Golden Chersonese And The Way Thither,* G. P. Putnam's & Sons, New York.

Prologue

1. Sykes, E. C. (1898) *Through Persia on a Side-Saddle.* A. D. Innes & Company, Ltd., London.
2. Ibid.
3. Ellis, Mrs S. (1839) *Women of England.* Fisher, Son & Co., London.
4. Marsden, K. (1893) *On Sledge and Horseback to Outcast Siberian Lepers.* The Record Press Ltd., London.
5. Ibid.
6. Ibid.
7. Frederick, B., and McLeod, S. H., (Eds.) (1993) *Women and the Journey: The Female Travel Experience.* Washington, Washington State University Press.
8. Bird, I. L. (1886) *A Lady's Life in the Rocky Mountains.* G. P. Putnam's & Sons, London and New York.
9. Robinson, J. (1991) *Wayward Women: A guide to women travellers.* Oxford University Press, Oxford.
10. Zheutlin, P. (2008) *Around the World on Two Wheels: Annie Londonderry's Extraordinary Ride.* New York, Citadel Press.
11. *The World (New York),* 3 July 1894.
12. Zheutlin, P. (2008) Around the World on Two Wheels: Annie Londonderry's Extraordinary Ride. New York, Citadel Press.
13. Campbell Davidson, L. (1889) *Hints to Lady Travellers: At Home and Abroad.* Illife & Son, London.

14. Ellis, K. (2008) *Star of the Morning: The Extraordinary Life of Lady Hester Stanhope*. Harper Collins, London.
15. Ibid.
16. Robinson, J. (2001) *Unsuitable for Ladies: An Anthology of Women Travellers.* Oxford University Press, Oxford.
17. French Sheldon, M. (1892) *Sultan to Sultan: Adventures among the Masai and other Tribes of East Africa.* Saxon & Co., London.
18. Ibid.
19. Ibid.
20. Ibid.
21. Ibid.
22. Ibid.
23. Sykes, E. C. (1898) *Through Persia on a Side-Saddle*. A. D. Innes & Company, Ltd., London.
24. Ibid.
25. Burlend, R. [Anon] (1848) *A True Picture of Emigration: or Fourteen Years in the Interior of North America*. G. Berger, London.
26. Ibid.
27. Hughes, Mrs T. F (1881) *Among the Sons of Han: Notes of a Six Years Residence in Various Parts of China and Formosa.* Tinsley Brothers, London.
28. Pfeiffer, I. (1852) *Visit to the Holy Land, Egypt and Italy*. Ingram, Cooke., London.
29. Livingstone, W. P. (1923) *Mary Slessor of Calabar: Pioneer Missionary*. Hodder & Stoughton, London.
30. Ibid.
31. Kingsley, M. H. (1901) *West African Studies*. Macmillan & Co. Ltd., London.
32. Livingstone, W. P. (1923) *Mary Slessor of Calabar: Pioneer Missionary*. Hodder & Stoughton, London.
33. Ibid.
34. Kingsley, M. H. (1901) *West African Studies*. Macmillan & Co. Ltd., London.

35. Mary Kingsley to John Holt, 20 February 1899. Quoted in Frank, K. (1986) *A Voyager Out: The Life of Mary Kingsley.* Houghton Mifflin Company, Boston.
36. Mary Kingsley to John Holt, 18 September 1899. Quoted in Flint, J. E., 'Mary Kingsley—a reassessment', *Journal of African History*, lv, i (1963), pp. 95-104.
37. Kingsley, M. H. (1901) *West African Studies.* Macmillan & Co. Ltd., London.
38. Ibid.
39. Ibid.
40. Ibid.
41. Ibid.

Part 1

1. Rijnhart, Dr S. C. (1904) *With the Tibetans in Tent and Temple: Narrative of four years' residence on the Tibetan border, and of a journey into the far interior.* Fleming H. Revell Co., London.
2. Ibid.
3. Ibid.
4. *The Pilgrimage of Etheria* (1919) Translated and annotated by Mrs M. L. McClure and C. L. Feltoe, Society For Promoting Christian Knowledge, London.
5. Ibid.
6. Bronte, C. (1847) *Jane Eyre.* Penguin edition (pub. 1996) Penguin Books, London.
7. Ibid.
8. *Burton Chronicle,* 3 September 1896.
9. *Aberdeen Press and Journal*, 28 December 1894.
10. *Manchester Courier and Lancashire General Advertiser*, 27 December 1894.
11. *Manchester Courier and Lancashire General Advertiser*, 27 July 1889.
12. Ibid.
13. Robinson, J. (1991) *Wayward Women: A guide to women travellers.* Oxford University Press, Oxford.

14. *Toronto Daily Mail,* 13 April 1892.
15. Ibid.
16. *The Young Woman,* 3 April 1896.
17. Ibid.
18. Ibid.
19. Ibid.
20. Rijnhart, Dr S. C. (1904) *With the Tibetans in Tent and Temple: Narrative of four years' residence on the Tibetan border, and of a journey into the far interior.* Fleming H. Revell Co., London.
21. Ibid.
22. Ibid.
23. Ibid.
24. Ibid.
25. Ibid.
26. Ibid.
27. Ibid.
28. Ibid.
29. Ibid.
30. Ibid.
31. Ibid.
32. Hore, A. (1886) *To Lake Tanganyika in a Bath Chair.* Sampson Low, Marston, Searle and Rivington, London.
33. Ibid.
34. Ibid.
35. Ibid.
36. Ibid.
37. *Pall Mall Gazette,* 30 December 1886.
38. Ibid.
39. *The Field, The Country Gentleman's Newspaper*, 18 December 1886.
40. Ibid.
41. Ibid.
42. Ibid.
43. Ibid.
44. Taylor, A. R. (1895) *Pioneering in Tibet.* Morgan and Scott, London.

45. Ibid.
46. Ibid.
47. Ibid.
48. Ibid.
49. Taylor, A. R. (1894) *The Origin of the Tibetan Pioneer Mission, Together with Some Facts about Tibet.* Morgan and Scott, London.
50. Ibid.
51. Ibid.
52. Ibid.
53. Ibid.
54. Ibid.
55. Robinson, J. (1991) *Wayward Women: A guide to women travellers.* Oxford University Press, Oxford.
56. Middleton, D. (1993) *Victorian Lady Travellers.* Academy Chicago Publishers, Chicago.
57. 'A Lady's Adventures in Tibet,' from *China's Millions magazine,* December 1893.
58. Ibid.
59. Ibid.

Part 2

1. *Daily Mirror,* 18 April 1978.
2. Fountaine, M. (1980) *Love Among the Butterflies: The Travels and Adventures of a Victorian Lady (Edited by W. F. Cater).* William Collins Sons & Co., London.
3. Ibid.
4. Ibid.
5. Ibid.
6. Ibid.
7. Ibid.
8. Ibid.
9. Ibid.
10. Ibid.
11. Ibid.
12. Ibid.

13. Morris, M. with O'Connor, L. (Eds.) (2007) *The Illustrated Virago Book of Women Travellers*. Virago Press, London.
14. Ibid.
15. Waring, S. (2015) 'Margaret Fountaine: a lepidopterist remembered.' Cited in *The Royal Society Journal of the History of Science*. Vol 69, Issue 1, pp53-68.
16. Ibid.
17. Fountaine, M. (1980) *Love Among the Butterflies: The Travels and Adventures of a Victorian Lady (Edited by W. F. Cater)*. William Collins Sons & Co., London.
18. Ibid.
19. Ibid.
20. Waring, S. (2015) 'Margaret Fountaine: a lepidopterist remembered.' *The Royal Society Journal of the History of Science*. Vol 69, Issue 1, pp53-68.
21. Irwin, T. (1978) 'Of butterflies and broken hearts.' *Antenna2*, Vol. 72. Cited in Waring, S. (2015) 'Margaret Fountaine: a lepidopterist remembered.' *The Royal Society Journal of the History of Science*. Vol 69, Issue 1, pp53-68.
22. Riley, N. (1940) 'Unpublished obituary of Margaret Fountaine' in the Fountaine Papers, Castle Museum, Norwich. Cited in Waring, S. (2015) 'Margaret Fountaine: a lepidopterist remembered.' *The Royal Society Journal of the History of Science*. Vol 69, Issue 1, pp53-68.
23. *Newry Telegraph,* 16 March 1830.
24. *Waterford Mail,* 10 April 1830.
25. *The Star*, 1 April 1873.
26. Ibid.
27. *South London Chronicle,* 2 August 1873.
28. Ibid.
29. *Birmingham Mail*, 30 September 1881.
30. Ibid.
31. Meryon, Dr C. (1845) *Memoirs of Lady Hester Stanhope.* Henry Colburn, London.
32. Ibid.
33. Cleveland, Duchess of (1914) *The Life and Letters of Lady Hester Stanhope.* John Murray, London.

34. Ibid.
35. Personal correspondence of Dr Charles Meryon, 23 March 1810. Held at the Centre for Kentish Studies, U1590 S6/2.
36. *Barbados Mercury and Bridgetown Gazette,* 6 July 1816.
37. *Westmoreland Gazette,* 19 October 1839.
38. *Leeds Intelligencer,* 19 October 1839.
39. Russell, M. (1994) *The Blessings of a Good Thick Skirt: Women Travellers and Their World.* Flamingo, London.
40. Davenport Adams, W. H. (1903) *Celebrated Women Travellers of the Nineteenth-Century.* E. P. Dutton & Co., New York.
41. *Belfast Commercial Chronicle,* 7 January 1839.
42. Ibid.
43. Ibid.
44. Ibid.
45. Ellis, K. (2008) *Star of the Morning: The Extraordinary Life of Lady Hester Stanhope.* Harper Collins, London.
46. *Leeds Intelligencer,* 19 October 1839.
47. *Hampshire Telegraph,* 29 August 1836.
48. *Letter from Arthur Phillip to the Marquis of Lansdowne, 3 July 1788.* Mitchell Library, State Library of New South Wales, MLMSS 7241.
49. McGoogan, K (2005) *Lady Franklin's Revenge: A True Story of Ambition, Obsession and the Remaking of Arctic History.* Harper Collins, Canada.
50. *Kilkenny Moderator,* 21 June 1837.
51. *Fife Herald,* 25 October 1838.
52. Woodward, Frances, J. 'Franklin, Lady Jane (1791–1875)', *Australian Dictionary of Biography,* National Centre of Biography, Australian National University
53. *Freeman's Journal,* 27 June 1838. 156
54. Ibid.
55. *London Evening Standard,* 15 June 1844.
56. *Illustrated London News,* 24 May 1845.
57. Ibid.
58. *Nottinghamshire Guardian,* 7 March 1850.
59. Undelivered letter from Lady Jane Franklin to her husband Sir John Franklin dated 30 Mar 1853. National Maritime Museum, Greenwich, London. (Archive Ref. AGC/F/9/1-3).

60. McGoogan, K (2005) *Lady Franklin's Revenge: A True Story of Ambition, Obsession and the Remaking of Arctic History*. Harper Collins, Canada.
61. *Evening Mail,* 30 May 1860.
62. Royal Geographical Society of London, 'Presentation of the Gold Medals to Lady Franklin and to Captain Sir F. L. Mc'Clintock.' Proceedings of the Royal Geographical Society of London, 28 May 1860.
63. *Saunders's Newsletter,* 19 November 1866.
64. *Morning Herald (London),* 16 November 1866.
65. *Illustrated London News,* 24 July 1875.
66. Ibid.
67. Bird, Isabella Lucy (1875) *The Hawaiian Archipelago: Six Months Among The Palm Groves, Coral Reefs, And Volcanoes Of The Sandwich Islands.* John Murray, London.
68. Ibid.
69. Campbell Davidson, L. (1889) *Hints to Lady Travellers: At Home and Abroad.* Illife & Son, London.
70. Bird, I. L. (1879) *A Lady's Life in the Rocky Mountains.* John Murray, London.
71. Ibid.
72. Ibid.
73. Ibid.
74. *The Belfast Newsletter,* 19 December 1879.
75. Bird, I. L. (1879) *A Lady's Life in the Rocky Mountains.* John Murray, London.
76. Ibid.
77. Ibid.

Part 3

1. Brodrick, M. (1900) *A handbook for travellers in Lower and Upper Egypt.* John Murrray, London.
2. *Egypt Papers*, vii–viii. Cited in Gill, David. "Brodrick, Mary (1858–1933), archaeologist." Oxford Dictionary of National Biography. 23 September 2004.

3. Ibid.
4. Robinson, E. (1937) Eds., *Egypt: Papers & Lectures by the late May Brodrick*. Alexander Moring Limited, London. Cited in Thornton, A. 'The Women Who Did.' *Archaeologists in Print*, 48–74. UCL Press, 2018.
5. Ibid.
6. *Egypt Papers*, vii–viii. Cited in Gill, David. "Brodrick, Mary (1858–1933), archaeologist." Oxford Dictionary of National Biography. 23 September 2004.
7. *Eastern Daily Press*, 5 October 1889.
8. Ibid.
9. Brodrick, M. (1900) *A handbook for travellers in Lower and Upper Egypt*. John Murrary, London.
10. *Penrith Observer*, 27 December 1898.
11. *Daily Mail*, 26 November 1906.
12. *Egypt Papers*, vii–viii. Cited in Gill, David. "Brodrick, Mary (1858–1933), archaeologist." Oxford Dictionary of National Biography. 23 September 2004.
13. *The Times,* 4 July 1890. Cited in Gill, David. "Brodrick, Mary (1858–1933), archaeologist." Oxford Dictionary of National Biography. 23 September 2004.
14. *Eastern Daily Press*, 26 November 1902.
15. *Exeter and Plymouth Gazette,* 10 November 1933.
16. Gill, David. "Brodrick, Mary (1858–1933), archaeologist." Oxford Dictionary of National Biography. 23 September 2004. 157
17. *Hastings and St Leonards Observer,* 9 August 1879.
18. *The Graphic*, 12 January 1878.
19. Ibid.
20. North, M. (1894) (Eds.) Symonds, J. C. *Recollection and Further Recollection of a Happy Life: being the Autobiography of Marianne North*. Vol. 1. New York, MacMillan.
21. Ibid.
22. Ibid.
23. Ibid.
24. Ibid.

25. Ibid.
26. North, M. (1894) (Eds.) Symonds, J. C. *Recollection and Further Recollection of a Happy Life: being the Autobiography of Marianne North.* Vol. 1. New York, MacMillan.
27. Ibid.
28. Ibid.
29. Ibid.
30. *Leeds Times,* 19 August 1882.
31. *Birmingham Post,* 24 June 1882.
32. *Hastings and St Leonards Observer,* 9 May 1885.
33. North, M. (1980) *A Vision of Eden: The Life and Work of Marianne North.* Webb & Bower, Exeter.

Part 4

1. Zheutlin, P. (2008) *Around the World on Two Wheels: Annie Londonderry's Extraordinary Ride.* New York, Citadel Press.
2. Ibid.
3. *The Northern Daily Mail,* 21 October 1895.
4. Ibid.
5. *Penny Illustrated Paper,* 6 July 1895.
6. *Singapore Straits Times,* 14 February 1895.
7. Ibid.
8. *The Northern Daily Mail,* 21 October 1895.
9. *San Francisco Chronicle,* 24 March 1895.
10. *Cycling Life,* March 1895.
11. Zheutlin, P. (2008) *Around the World on Two Wheels: Annie Londonderry's Extraordinary Ride.* New York, Citadel Press.
12. Shadwell, A. (M.D.) 'The Hidden Dangers of Cycling' in *The National Review*, 1 February 1897.
13. Ibid.
14. *The Phrenological Journal and Science of Health, Volume 104, 1897.* Fowler & Wells, New York.
15. Campbell Davidson, L. (1889) *Hints to Lady Travellers: At Home and Abroad.* Illife & Son, London.

16. Campbell Davidson, L. (1896) *Handbook for Lady Cyclists*. Hay Nisbett & Co, London.
17. *New York Sunday World*, 20 October 1895.
18. *The New York World*, 6 October 1889.
19. Bly, N. (1888) *Six Months in Mexico*. American Publishers Corporation, New York.
20. Ibid.
21. Ibid.
22. Ibid.
23. Ibid.
24. Ibid.
25. Bly, N. (1890) *Around the World in Seventy-Two Days*. Pictorial Weeklies, New York.
26. Ibid.
27. Ibid.
28. *The Luton Reporter and Beds. and Herts. News,* 14 December 1889.
29. *The Leeds Times,* 1 February 1890.
30. Ibid. 158
31. *Irish Society,* 28 December 1889.
32. Ibid.
33. Bly, N. (1890) *Around the World in Seventy-Two Days*. Pictorial Weeklies, New York.
34. Ibid.
35. Goodman, M. (2013) *Eighty Days: Nellie Bly and Elizabeth Bisland's History-Making Race Around the World.* Ballantine Books, New York.
36. Bly, N. (1890) *Around the World in Seventy-Two Days*. Pictorial Weeklies, New York.
37. Ibid.
38. Ibid.
39. Ibid.
40. Ibid.
41. Ibid
42. Ibid.
43. Ibid.
44. Ibid.

45. Ibid.
46. Ibid.
47. Ibid.
48. Ibid.
49. *The Bridport News,* 31 January 1890.
50. Bly, N. (1890) *Around the World in Seventy-Two Days.* Pictorial Weeklies, New York.
51. Ibid.
52. *Paisley & Renfrewshire Gazette*, 15 February 1890.
53. *Blackburn Standard,* 1 February 1890.
54. Campbell Davidson, L. (1889) *Hints to Lady Travellers: At Home and Abroad.* Illife & Son, London.
55. Bly, N. (1890) *Around the World in Seventy-Two Days.* Pictorial Weeklies, New York.
56. Ibid.
57. Ibid.
58. Ibid.
59. Ibid.
60. Ibid.
61. *New York World,* 26 January 1896.
62. Ibid.
63. Ibid.
64. Ibid.
65. Bankston, J. (2011) *Nellie Bly: Journalist.* Chelsea House, New York.
66. *Denver Post,* 26 December 1914.
67. *The Scotsman*, 19 May 1915.
68. Kroeger, B. (1994) *Nellie Bly: Daredevil, Reporter, Feminist.* Random House, New York. Cited in Bankston, J. (2011) *Nellie Bly: Journalist.* Chelsea House, New York.
69. *New York Evening Journal*, 8 January 1922.
70. Black, M. L. (1976) *Martha Black: Her story from the Dawson Gold Fields to the Halls of Parliament.* PR Services Ltd., Yukon, Canada.
71. Ibid.
72. Ibid.

73. Ibid.
74. Ibid.
75. Ibid.
76. Ibid.
77. Ibid.
78. Ibid.
79. Ibid.
80. Ibid.
81. Ibid.

Part 5

1. Moodie, S. (1871) *Rouging it in the Bush; or Forest Life in Canada.* Hunter, Rose and Company., Toronto.
2. Ibid.
3. Ibid.
4. *Perthshire Courier,* 3 March 1831.
5. Ibid.
6. Moodie, S. (1871) *Rouging it in the Bush; or Forest Life in Canada.* Hunter, Rose and Company., Toronto.
7. Ibid.
8. Ibid.
9. Ibid.
10. Ibid.
11. Ibid.
12. Ibid.
13. Ibid.
14. *Perthshire Courier,* 3 March 1831.
15. Moodie, S. (1871) *Rouging it in the Bush; or Forest Life in Canada.* Hunter, Rose and Company., Toronto.
16. Ibid.
17. Ibid.
18. Traill, C. P. (1836) *The Backwoods of Canada; being letters from the Wife of an Emigrant Officer, Illustrative of the Domestic Economy of British America.* Charles Knight, London.
19. Ibid.

20. Ibid.
21. Ibid.
22. Ibid.
23. Ibid.
24. Ibid.
25. Ibid.
26. Ibid
27. Ibid.
28. Ibid.
29. Moodie, S. (1871) *Rouging it in the Bush; or Forest Life in Canada.* Hunter, Rose and Company., Toronto.
30. Ibid.
31. Traill, C. P. (1854) *The Female Emigrants Guide and Hints on Canadian Housekeeping.* C.W., Toronto.
32. Ibid.
33. Ibid.
34. Ibid.
35. Ibid.
36. Ibid.
37. Ibid.
38. Bradley, E. (1823) *An Authentic Narrative of the Shipwreck and Sufferings of Mrs Eliza Bradley, The Wife of Capt. James Bradley, of Liverpool, Commander of the Ship Sally, which was Wrecked on the Coast of Barbary, in June, 1818.* Jonathan Howe, Boston.
39. Ibid.
40. Ibid.
41. Ibid.
42. Ibid.
43. Ibid.
44. Ibid.
45. Ibid.
46. Ibid.
47. Ibid.
48. Ibid.
49. Ibid.

50. Ibid.
51. Robinson, J. (1991) *Wayward Women: A guide to women travellers.* Oxford University Press, Oxford.
52. Bradley, E. (1823) *An Authentic Narrative of the Shipwreck and Sufferings of Mrs Eliza Bradley, The Wife of Capt. James Bradley, of Liverpool, Commander of the Ship Sally, which was Wrecked on the Coast of Barbary, in June, 1818.* Jonathan Howe, Boston.
53. Ibid.
54. Edgeworth David, C. (1899) *Funafuti Or Three Months On A Coral Island: An Unscientific Account Of A Scientific Expedition.* John Murray, London.
55. Ibid.
56. Ibid.
57. Ibid.
58. Ibid.
59. Gill, I. B. (1878) *Six Months in Ascension: An Unscientific Account of a Scientific Expedition.* John Murray, London.
60. Ibid.
61. Ibid.
62. Ibid.
63. Ibid.
64. Ibid.
65. Ibid.
66. Ibid.
67. Ibid.
68. Ibid.
69. Scammell, H. B. (1890) *Stanley and the white heroes in Africa; being an edition from Mr Stanley's late personal writings on the Emin Pasha relief expedition, etc.* Scammell & Company. St Louis and Philadelphia.
70. Ibid.
71. Middleton, D (2004) 'Baker, Florence Barbara Maria, Lady Baker (1841–1916)', Oxford Dictionary of National Biography. Oxford University Press, 2004.
72. *Knaresborough Post,* 13 December 1873.
73. *The Western Times,* 13 March 1916.

74. Schapera, I. (1960) (Eds.) *Livingstone's Private Journals.* Chatto & Windus, London.
75. Livingstone, D. (1857) *Missionary Travels and Researches in South Africa including a Sketch of Sixteen Years Residence in the Interior of Africa.* Ward, Lock & Co. Limited, London.
76. Ibid.
77. Ibid.
78. Schapera, I. (1960) (Eds.) *Livingstone's Private Journals.* Chatto & Windus, London.
79. Ibid.
80. Ibid.
81. Wallis, J. P. R. (1956) (Eds.) *The Zambezi Expedition of David Livingstone, 1858-1863.* Chatto & Windus Limited, London.
82. Ibid.
83. Ibid.
84. Blaikie, W. G. (1910) *The Personal Life of David Livingstone.* John Murray, London.
85. Livingstone, David, 1813-1873. 'Letter to Robert Moffat 1, 28, 29 April 1862.' Livingstone Online: livingstoneonline.org. Accessed: 18 May 2022.

Part 6

1. Crocker, A. (1936) *And I'd Do It Again.* Coward, McCann, Inc. New York.
2. Pfeiffer, Ida (1861) *The Last Travels of Ida Pfeiffer, inclusive a visit to Madagaskar.* Routledge, Warne and Routledge, London.
3. Pfeiffer, Ida (1852) *Visit to Iceland and the Scandinavian North.* Ingram, London.
4. Pfeiffer, I. (1852) *Visit to the Holy Land, Egypt and Italy.* Ingram, Cook & Co., London. 161
5. Ibid..
6. Ibid..
7. Ibid..
8. Ibid..

9. Ibid.
10. Ibid.
11. Ibid.
12. Ibid.
13. Pfeiffer, I. (1852) *A journey to Iceland and travels in Sweden and Norway*. George P. Putnam, New York.
14. Ibid.
15. *Morning Herald*, 28 October 1852.
16. Pfeiffer, I. (1852) *A journey to Iceland and travels in Sweden and Norway*. George P. Putnam, New York.
17. Ibid.
18. Ibid.
19. Ibid.
20. *The Globe*, 24 August 1852.
21. Pfeiffer, I. (1850) *A Woman's Journey Round The World from Vienna to Brazil, Chile, Tahiti, China, Hindostan, Persia and Asia Minor*. Office of the National Illustrated Library, London.
22. Ibid.
23. Ibid.
24. Ibid.
25. Ibid.
26. Ibid.
27. Ibid.
28. Ibid.
29. Ibid.
30. Ibid.
31. Ibid.
32. Ibid.
33. Ibid.
34. Ibid.
35. Ibid.
36. Ibid.
37. Ibid.
38. Ibid.
39. Ibid.

Notes

40. Pfeiffer, I. (1855) *A Lady's Second Journey Round the World: From London to the Cape of Good Hope, Borneo, Java, Sumatra, Celebes, Ceram, The Molucca's etc. California, Panama, Peru, Ecuador, and the United States. Vol. 1.* Longman, Brown, Green and Longmans, London.
41. Ibid.
42. Ibid.
43. *Lloyd's Weekly Newspaper*, 31 December 1854.
44. Ibid.
45. Ibid.
46. Poole, S. (1851) *The Englishwoman in Egypt: Letters from Cairo Written during a Residence There in 1842, 3, & 4 with E. W. Lane Esq.* 3 vols. 1844-6. C. Cox, London.
47. Ibid.
48. Pfeiffer, I. (1850) *A Woman's Journey Round The World from Vienna to Brazil, Chile, Tahiti, China, Hindostan, Persia and Asia Minor.* Office of the National Illustrated Library, London.
49. *Home News for India, China and the Colonies,* 24 November 1851.
50. Ibid.
51. Ibid.
52. Davenport Adams, W. H. (1903) *Celebrated Women Travellers of the Nineteenth Century.* E. P. Dutton & Co., New York. 162
53. Crocker, A. (1936) *And I'd Do It Again.* Coward, McCann, Inc. New York.
54. *Totnes Weekly Times,* 22 September 1894.
55. *San Francisco Examiner* syndicated in the *Clarion*, 30 December 1899.
56. *Buckingham Advertiser and Free Press,* 20 July 1889.
57. Crocker, A. (1936) *And I'd Do It Again.* Coward, McCann, Inc. New York.
58. Ibid.
59. Ibid.
60. Ibid.
61. *Cambridge Independent Press,* 27 January 1883.
62. *Western Gazette,* 26 January 1883.

63. Crocker, A. (1936) *And I'd Do It Again.* Coward, McCann, Inc. New York.
64. Ibid.
65. Ibid.
66. Ibid.
67. Ibid.
68. Ibid.
69. Ibid.
70. Ibid.
71. Ibid.
72. Ibid.
73. Ibid.
74. Ibid.
75. Ibid.
76. Ibid.
77. *The Graphic,* 7 May 1927.
78. Crocker Gouraud, A. (1910) *Moon-Madness and Other Fantasies.*
79. Kevin Taylor. *Who Is Aimee Crocker?* [Online]
80. *Brooklyn Daily Eagle,* 8 February 1941.
81. *Oakland Tribune,* 17 January 1926.
82. *American Weekly,* 10 May 1942.
83. Ibid.

Epilogue

1. Bird, I. L. (1883) *The Golden Chersonese And The Way Thither,* G. P. Putnam's & Sons, New York.
2. Sheldon, M. F. (1892) *Sultan to Sultan: Adventures among the Masai and other Tribes of East Africa.* Saxon & Co., London.
3. Ibid.
4. Ibid.
5. Eyre, M. (1865) *A Lady's Walks in the South of France in 1863.* Richard Bentley, London.
6. Kingsley, M. (1897) 'West Africa from an Ethnologist's Point of View.' *Transactions of the Liverpool Geographical Society*, pp. 58–73.

7. Cater, W. F. (1988) (Eds.) Fountaine, M. *Butterflies and Late Loves. The Further Travels and Adventures of a Victorian Lady*. Penguin, London.
8. Foster, M. and Foster, M. (1987) *Forbidden Journey: The Life of Alexandra David-Neel*. Harper & Row, San Francisco.
9. Ibid. 163

Bibliography

Books

Bankston, J. (2011) *Nellie Bly: Journalist.* Chelsea House, New York.

Bird, I. L. (1875) *The Hawaiian Archipelago: Six Months Among The Palm Groves, Coral Reefs, And Volcanoes Of The Sandwich Islands.* John Murray, London.

Bird, I. L. (1883) *The Golden Chersonese And The Way Thither*, G. P. Putnam's & Sons, New York.

Bird, I. L. (1879) *A Lady's Life in the Rocky Mountains.* John Murray, London.

Bird, I. L. (1886) *A Lady's Life in the Rocky Mountains.* G. P. Putnam's & Sons, London and New York.

Black, M. L. (1976) *Martha Black: Her story from the Dawson Gold Fields to the Halls of Parliament.* PR Services Ltd., Yukon, Canada.

Blaikie, W. G. (1910) *The Personal Life of David Livingstone.* John Murray, London.

Bly, N. (1890) *Around the World in Seventy-Two Days.* Pictorial Weeklies, New York.

Bradley, E. (1823) *An Authentic Narrative of the Shipwreck and Sufferings of Mrs Eliza Bradley, The Wife of Capt. James Bradley, of Liverpool, Commander of the Ship Sally, which was Wrecked on the Coast of Barbary, in June, 1818.* Jonathan Howe, Boston.

Brodrick, M. (1900) *A handbook for travellers in Lower and Upper Egypt.* John Murrary, London.

Bronte, C. (1847) *Jane Eyre.* Penguin edition (pub. 1996) Penguin Books, London.

Burlend, R. [Anon] (1848) *A True Picture of Emigration: or Fourteen Years in the Interior of North America.* G. Berger, London.

Bibliography

Campbell Davidson, L. (1889) *Hints to Lady Travellers: At Home and Abroad*. Illife & Son, London.

Campbell Davidson, L. (1896) *Handbook for Lady Cyclists*. Hay Nisbett & Co, London.

Cater, W. F. (1988) (Eds.) Fountaine, M. *Butterflies and Late Loves. The Further Travels and Adventures of a Victorian Lady*. Penguin, London.

Crocker, A. (1936) *And I'd Do It Again*. Coward, McCann, Inc. New York.

Crocker Gouraud, A. (1910) *Moon-Madness and Other Fantasies*.

Davenport Adams, W. H. (1903) *Celebrated Women Travellers of the Nineteenth Century*. E. P. Dutton & Co., New York.

David Neel, A. (1927) *My Journey to Lhasa: The Personal Story of the only White Woman Who Succeeded in Entering the Forbidden City*. William Heinemann Ltd., London.

Dixie, Lady F. (1880) *Across Patagonia*. Richard Bentley & Son, London.

Duncan, J. E. (1906) *A Summer Ride Through Western Tibet*. Smith, Elder & Co., London.

Edgeworth David, C. (1899) *Funafuti Or Three Months On A Coral Island: An Unscientific Account Of A Scientific Expedition*. John Murray, London.

Ellis, K. (2008) *Star of the Morning: The Extraordinary Life of Lady Hester Stanhope*. Harper Collins, London.

Ellis, Mrs S. (1839) *Women of England*. Fisher, Son & Co., London.

Eyre, M. (1865) *A Lady's Walks in the South of France in 1863*. Richard Bentley, London.

Fountaine, M. (1980) *Love Among the Butterflies: The Travels and Adventures of a Victorian Lady (Edited by W. F. Cater)*. William Collins Sons & Co., London.

Foster, M. and Foster, M. (1987) *Forbidden Journey: The Life of Alexandra David-Neel*. Harper & Row, San Francisco.

Frank, K. (1986) *A Voyager Out: The Life of Mary Kingsley*. Houghton Mifflin Company, Boston

French Sheldon, M. (1892) *Sultan to Sultan: Adventures among the Masai and other tribes of East Africa*. Arena Publishing Co., Massachusetts.

Frederick, B., and McLeod, S. H., (Eds.) (1993) *Women and the Journey: The Female Travel Experience*. Washington, Washington State University Press.

Gill, I. B. (1878) *Six Months in Ascension: An Unscientific Account of a Scientific Expedition*. John Murray, London. 164

Goodman, M. (2013) *Eighty Days: Nellie Bly and Elizabeth Bisland's History-Making Race Around the World*. Ballantine Books, New York.

Gribbin, M. & Gribbin, J. (2008) *Flower Hunters*. Oxford University Press, Oxford.

Hore, A. (1886) *To Lake Tanganyika in a Bath Chair*. Sampson Low, Marston, Searle and Rivington, London.

Hughes, Mrs T. F (1881) *Among the Sons of Han: Notes of a Six Years Residence in Various Parts of China and Formosa*. Tinsley Brothers, London.

Kabbani, R. (1988) (Eds.) *The Passionate Nomad: The Diary of Isabelle Eberhardt*. Beacon Press, Boston.

Kingsley, M. H. (1897) *Travels in West Africa*. Macmillan & Co., London.

Kingsley, M. H. (1901) *West African Studies*. Macmillan & Co. Ltd., London.

Kroeger, B. (1994) *Nellie Bly: Daredevil, Reporter, Feminist*. Random House, New York.

Livingstone, D. (1857) *Missionary Travels and Researches in South Africa including a Sketch of Sixteen Years Residence in the Interior of Africa*. Ward, Lock & Co. Limited, London.

Livingstone, W. P. (1923) *Mary Slessor of Calabar: Pioneer Missionary*. Hodder & Stoughton, London.

Lowe, E. (1857) *Unprotected Females in Norway; or the Pleasantest Way of Travelling There, Passing Through Denmark and Sweden with Scandinavian Sketches from Nature*. G. Routledge & Co., London.

Marsden, K. (1893) *On Sledge and Horseback to Outcast Siberian Lepers*. The Record Press Ltd., London.

Mazuchelli, N. E. (1876) *The Indian Alps and How We Crossed Them*. Longmans, Green, & Co., London.

Bibliography

McGoogan, K (2005) *Lady Franklin's Revenge: A True Story of Ambition, Obsession and the Remaking of Arctic History*. Harper Collins, Canada.

Meakin, A. (1901) *A Ribbon of Iron*. Archibald Constable & Co., London.

Meredith, Mrs C. (1852) *My Home in Tasmania, During a Residence of Nine Years: In Two Volumes*. John Murray, London.

Meryon, Dr C. (1845) *Memoirs of Lady Hester Stanhope*. Henry Colburn, London. Cited in Ellis, K. (2008) *Star of the Morning: The Extraordinary Life of Lady Hester Stanhope*. Harper Collins, London.

Middleton, D. (1993) *Victorian Lady Travellers*. Academy Chicago Publishers, Chicago.

Moodie, S. (1871) *Rouging it in the Bush; or Forest Life in Canada*. Hunter, Rose and Company., Toronto.

Morris, M. with O'Connor, L. (Eds.) (2007) *The Illustrated Virago Book of Women Travellers*. Virago Press, London.

North, M. (1894) (Eds.) Symonds, J. C. *Recollections and Further Recollections of a Happy Life: being the Autobiography of Marianne North*. Vol. 1. New York, MacMillan.

North, M. (1980) *A Vision of Eden: The Life and Work of Marianne North*. Webb & Bower, Exeter.

Pfeiffer, I. (1850) *A Woman's Journey Round The World from Vienna to Brazil, Chile, Tahiti, China, Hindostan, Persia and Asia Minor*. Office of the National Illustrated Library, London.

Pfeiffer, I. (1852) *Visit to the Holy Land, Egypt and Italy*. Ingram, Cooke., London.

Pfeiffer, Ida (1852) *Visit to Iceland and the Scandinavian North*. Ingram, London.

Pfeiffer, I. (1852) *A journey to Iceland and travels in Sweden and Norway*. George P. Putnam, New York.

Pfeiffer, I. (1855) *A Lady's Second Journey Round the World: From London to the Cape of Good Hope, Borneo, Java, Sumatra, Celebes, Ceram, The Molucca's etc. California, Panama, Peru, Ecuador, and the United States. Vol. 1*. Longman, Brown, Green and Longmans, London.

Pfeiffer, Ida (1861) *The Last Travels of Ida Pfeiffer, inclusive a visit to Madagaskar.* Routledge, Warne and Routledge, London.

Poole, S. (1851) *The Englishwoman in Egypt: Letters from Cairo Written during a Residence There in 1842, 3, & 4 with E. W. Lane Esq.* 3 vols. 1844-6. C. Cox, London.

Rijnhart, Dr S. C. (1904) *With the Tibetans in Tent and Temple: Narrative of four years' residence on the Tibetan border, and of a journey into the far interior.* Fleming H. Revell Co., London.

Robinson, E. (1937) (Eds.) *Egypt: Papers & Lectures by the late May Brodrick.* Alexander Moring Limited, London.

Robinson, J. (1991) *Wayward Women: A guide to women travellers.* Oxford University Press, Oxford. 165

Robinson, J. (2001) *Unsuitable for Ladies: An Anthology of Women Travellers.* Oxford University Press, Oxford.

Russell, M. (1994) *The Blessings of a Good Thick Skirt: Women Travellers and Their World.* Flamingo, London.

Scammell, H. B. (1890) *Stanley and the white heroes in Africa; being an edition from Mr Stanley's late personal writings on the Emin Pasha relief expedition, etc.* Scammell & Company. St Louis and Philadelphia.

Schapera, I. (1960) (Eds.) *Livingstone's Private Journals.* Chatto & Windus, London.

Sykes, E. C. (1898) *Through Persia on a Side-Saddle.* A. D. Innes & Company, Ltd., London.

Taylor, A. R. (1894) *The Origin of the Tibetan Pioneer Mission, Together with Some Facts about Tibet.* Morgan and Scott, London.

Taylor, A. R. (1895) *Pioneering in Tibet.* Morgan and Scott, London.

The Pilgrimage of Etheria (1919) Translated and annotated by Mrs M. L. McClure and C. L. Feltoe, Society For Promoting Christian Knowledge, London.

Traill, C. P. (1836) *The Backwoods of Canada; being letters from the Wife of an Emigrant Officer, Illustrative of the Domestic Economy of British America.* Charles Knight, London.

Traill, C. P. (1854) *The Female Emigrants Guide and Hints on Canadian Housekeeping.* C.W., Toronto.

Wallis, J. P. R. (1956) (Eds.) *The Zambezi Expedition of David Livingstone, 1858-1863*. Chatto & Windus Limited, London.

Zheutlin, P. (2008) *Around the World on Two Wheels: Annie Londonderry's Extraordinary Ride*. New York, Citadel Press.

Journals

Irwin, T. (1978) 'Of butterflies and broken hearts.' *Antenna2*, Vol. 72.

Mary Kingsley to John Holt, 18 September 1899. Quoted in Flint, J. E., 'Mary Kingsley—a reassessment', *Journal of African History*, lv, i (1963), pp. 95-104.

Kingsley, M. (1897) 'West Africa from an Ethnologist's Point of View.' *From Transactions of the Liverpool Geographical Society*, pp. 58–73.

'Monumental Commemorations'. *Art-Journal*, Vol. 5, Issue 11, pp37-38.

Shadwell, A. (M.D.) 'The Hidden Dangers of Cycling' in *The National Review*, 1 February 1897.

The Phrenological Journal and Science of Health, Volume 104, 1897. Fowler & Wells, New York.

Thornton, A. 'The Women Who Did.' *Archaeologists in Print*, 48–74. UCL Press, 2018.

Waring, S. (2015) 'Margaret Fountaine: a lepidopterist remembered.' *The Royal Society Journal of the History of Science*. Vol 69, Issue 1, pp53-68.

Newspapers and Magazines

Aberdeen Press and Journal
American Weekly
Barbados Mercury and Bridgetown Gazette
Belfast Commercial Chronicle
Birmingham Mail
Blackburn Standard
Buckingham Advertiser and Free Press

Burton Chronicle
Brooklyn Daily Eagle
Cambridge Independent Press
China's Millions
Clarion
Daily Chronicle
Daily Mail
Daily Mirror
Denver Post
Dunfermline Saturday Press 166
Eastern Daily Press
Evening Mail
Exeter and Plymouth Gazette
Fife Herald
Freeman's Journal
Hampshire Telegraph
Hastings and St Leonards Observer
Home News for India, China and the Colonies
Illustrated London News
Irish Society
Kilkenny Moderator
Knaresborough Post
Leeds Intelligencer
Lloyd's Weekly Newspaper
London Evening Standard
Manchester Courier and Lancashire General Advertiser
Morning Herald (London)
Newry Telegraph
New York Sunday World
New York Evening Journal
Northampton Mercury
Nottinghamshire Guardian
Oakland Tribune
Paisley & Renfrewshire Gazette
Pall Mall Gazette
Penny Illustrated Paper

Penrith Observer
Perthshire Courier
San Francisco Chronicle
San Francisco Examiner
Saunders's Newsletter
Singapore Straits Times
South London Chronicle
The Belfast Newsletter
The Bridport News
The Era
The Field, The Country Gentleman's Newspaper
The Glasgow Daily Herald
The Globe
The Graphic
The Luton Reporter and Beds. and Herts. News
The Leeds Times
The Northern Daily Mail
The Scotsman
The Star
The Times
The Western Times
The World (New York)
The Young Woman
Toronto Daily Mail
Totnes Weekly Times
Waterford Mail
Western Gazette
Westmoreland Gazette 167

Websites

AimeeCrocker.com
archive.org.uk
Australian Dictionary of Biography
British Newspaper Archive
National Maritime Museum

Oxford Dictionary of National Biography
Royal Society Publishing
Royal Geographical Society
University of Cambridge Scott Polar Research Institute

Archives

Updates and reports on the exploits of the most famous lady explorers of the day can be found in the British Newspaper Archive (accessed online at www.britishnewspaperarchive.co.uk), as well as their obituaries.

Digitally scanned copies of the original editions of their narratives can also be found online at archive.org.uk.

The location of collections relating to specific individuals can be established via Archives Hub (archiveshub.jisc.ac.uk). It primarily covers British archives, and whilst it does not show copies of digitised documents, it provides a link to the original archive and which institution holds it. You can then check individual institution websites for any documents they have digitised for public viewing online

Index

Baggage *see* luggage.
Baker, Florence, 136-38, 145, 164
   early life, 136
   death, 138
   journey to the source of the Nile, 136-38
   marriage, 137
   return to England, 138
   suppression of slave trade, 138
Bird, Isabella, xviii, xi, 62-7
   marriage, 67
   press reactions to travelogue, 65-6
   relationship with Jim Nugent, 63-6
   travel in the East, 66
Black, Martha, 105-10
   adolescence, 105-106
   birth, 105
   death, 110
   journey to the Klondike, 107-109
   political life, 110
   Life in Dawson, 108-110
Bly, Nellie, 88-105
   around the world trip, 91-9
   background in journalism and exposé, 88-91, 102
   British reactions, 92-3
   charitable acts, 104
   death, 104
   Elizabeth Bisland rival attempt, 93, 99
   meeting with Jules Verne, 96
   merchandise, 101-102
   opinions on the opposite sex, 101
   travels in Mexico, 89-91
   war journalism, 103-104
   women's suffrage, 102-103
Bradley, Eliza, 123-31
   birth, 123
   departs from England, 123
   capture, 126
   in captivity, 126-28
   return to England, 129
   similarities to fictious accounts, 129-31
Brodrick, Mary, 68-74, 164
   birth, 68
   British Museum lectures, 69, 71
   death, 73

education,
  in France, 69
  in England, 70
exploration of Egypt, 68, 70-2
translations, 71-2
Boyle Hore, Annie, 13-17, 164
  bath chair, 14, 17
  opinions on slavery, 15
  response of press, 15-16
  time in Kavala, 16-17
  travelling with son, 13-14, 16
Brontë, Charlotte:
  Jane Eyre, 1, 4
Burlend, Rebecca, xxv-xxvi

Campbell Davidson, Lillias, xv, xxiii, 12, 63, 87, 100
Chaucer, Geoffrey:
  The Wife of Bath, 3
Class, xiv, 17, 34, 65, 90, 128, 164
Colonial culture, xi, xxviii-xxix, 53, 156
Companion, xxv, 23, 57, 62-3, 76, 136, 154, 173
Crocker, Aimee, 146, 166-76
  assassination attempts, 166-67, 172-73
  birth and inheritance, 166
  Buddhism, 173-75, 167
  death, 175
  destruction of shrine, 175-76
  marriages and divorces, 167-69, 175
  romantic liaisons, 168, 171, 175

Dangers, xxvi-xxvii, 1, 19, 25, 42, 65, 139, 144
  bandits and robbers, xvi, 1, 4-5, 11-12, 16, 154
  climate, 4, 126, 142, 159,
  harassment, xv, 100-101, 158
  illness and disease, 4, 7, 16, 42, 132, 136, 148, 177
  indigenous people, 11-12, 19-21, 42, 125-27, 130
  poisoning, 20, 166, 173
  political situation, xx, 5, 8, 43, 90, 161
  pregnancy, 140-41, 143, 109, 141, 143
  violence, xv, 16, 154, 158
  terrain, xxiv-xxv, 126, 142, 151, 159
  transport, xxiii-xxv, 84, 95, 123-25, 148, 154, 168-69
  wild animals, xxiv, 1, 51-2, 157
David-Neel, Alexandra, x, 179
Digby el Mezrab, Lady Jane, 32-7
  birth, 32
  celebrity status, 33, 37
  death, 36
  departs England, 33
  coverage in the press and media, 33, 35, 37
  love affairs, 33-4
  marriages, 32, 34
  relationship with Isabel Burton, 34-5

Index

relationship with Sheik Medjuel el Mezrab, 34
married life, 34-5
Dixie, Lady Florence, xiii
Duncan, Jane Ellen, xvi

Eberhardt, Isabelle, xiv-xv
Edgeworth David, Cara, 131-32
life on Funafuti, 132
Edwards, Amelia, 68, 179
Etheria, 2-3
Eyre, Mary, 178

Faith, 3, 7, 12-3, 19, 21, 23, 34, 128, 140-43
see also missionary work
Fashion, xxii, 86, 89, 102, 128, 162
see also travelling attire
Femininity, xvi, xxi, 147, 149
Food, xiii, xviii, 9, 19-21, 113, 125-26, 153, 155
Fountaine, Margaret, 24-32
bequest, 24
birth, 25
early life, 25-7
death, 25
inheritance, 28
romantic endeavours, 26-31
Franklin, Lady Jane, 46-61
childhood and adolescence, 47
death, 61
life abroad:
in the Antipodes, 49-53
in the Mediterranean, 47-8
in later life, 58-9

marriage, 47
philanthropy, 51-2
reaction to death of Sir John Franklin, 57
return to England, 53-4
Royal Geographical Society Award, 57-8
search for Sir John Franklin, 55-7

Gill, Isobel, 132-35
death, 135
first impressions of Ascension, 133
life on Ascension, 133-35
philanthropy, 135
Guides, 1, 9-13, 20-1, 29, 52, 62, 64, 71, 142, 151, 156
see also Servants

Hughes, Mrs T. F., xxvi

Illness and disease *see* dangers
Imperialism, xxiv, xxviii-xxix
Indigenous people:
attitudes towards, xxix-xxx, 6, 17, 62, 95, 130, 155-56, 158
beliefs, xiv, 21
customs, xv, 19, 21, 70, 89, 91, 97, 112-14, 128, 132, 149, 156, 162, 172
exploitation of, xxiv, 34, 41, 43
rituals, 2, 21, 98, 149, 158

Kingsley, Mary, xii, xvii-xviii, xix, xxx, 179

213

Lane Poole, Sophia, 68, 162
Linnean Society, 31
Livingstone, Mary: 139-45
   birth, 139
   early life, 139
   death, 144
   first missionary posting, 140
   journey across the Kalahari, 141-43
   life in Kolobeng, 140-41
   loss of daughter, 141
   marriage, 139
   return to Africa, 144
   return to England, 143
Londonderry, Annie: xvi, 80-8
   becoming a 'New Woman', xxiii, 86-7
   'bicycle face', 86
   fashion advice, 86
   lecture tour, 84, 86
   return to America, 84-5
   the wager, xxii, 81
   touring the globe, 80, 82-4
Lowe, Emily, xiii
Luggage, xxiii, xv-xix, xxv, xxx, 39, 68, 92, 148

Marsden, Kate, xii, xix-xx, xxiii
Mazuchelli, Nina Elizabeth, xiii
Meakin, Annette, xv
Missionary work, xi, xiv, xxviii, 2-7, 13-23, 139-44, 164, 167, 170
   China Inland Mission, 18

London Missionary Society, 4, 14
   See also faith
Moodie, Susanna: 111-16
   birth, 112
   death, 120
   marriage, 112
   life on Kutchawanook Lake, 114-16
   life on Lake Ontario, 112-14
   press accounts of emigration, 112, 114
   thoughts on emigration, 120

'New woman', 86, 88, 166
   depictions in the press, 165-66
North, Marianne: 73-9, 164
   birth, 75
   early life, 75
   death, 79
   death of father, 75-6
   connection to Royal Botanical Gardens Kew, 74-5, 78
Nursing, xii, xix-xx, xxviii, 138, 179

Parr Traill, Catharine: 116-23, 165
   birth, 122
   death, 122
   exploration, 119
   thoughts on emigration, 117-18, 121-22
   published works, 116, 119, 121-22

Pfeiffer, Ida: xxvi, 146-65
  birth and early years, 146-47
  explorations of:
    Brazil and Chile, 153-56
    Ceylon and Madras, 157-58
    China and Hong Kong, 156-57
    Corfu, 159
    Iraq, 158-59
    Madagascar and attempted political coup, 161
    Russian Empire, 159
    Sweden, Norway and Iceland, 149-53
    Tahiti, 155-56
    the Holy Land and Egypt, xxvi, 147
  marriage and married life, 147
  opinions regarding indigenous peoples, 156, 160-62

Race,
  theories of, xxviii, 156
Rijnhart, Susie: xiv, 1, 7-13
  attempt on Lhasa, 9
  birth, 7
  birth of son, 8
  death, 13
  death of Petrus Rijnhart, 11, 13
  death of son, 1, 10
  education, 7
  in Lusar, 8
  marriages, 7, 13
  return to Canada, 13
  return to China, 13
Robertson, Janet, xxiii
Royal Geographical Society, xii, xxi, 57-8, 110, 138, 163
Royle Taylor, Annie: 12, 17-23
  arrest and expulsion from Tibet, 21-2
  attempted murder of, 19-20
  attempt on Lhasa, 19-21
  childhood and education, 17-18
  death, 22
  in China, 18-19
  missionary work in England, 18
  opinions on Tibetan beliefs, 19, 21
  press coverage, 20, 23
  relationship with Pontso, 20-2

Servants, 20-1, 33, 42-3, 46, 48, 76, 101, 113, 142, 158, 164, 167, 172
  see also Companions
  see also Guides
Sexual impropriety, 167, 172
Shaw, Flora, xxviii-xxix
Sheldon, May French, xiii, xxiv-xxv, 157, 177-78
Slessor, Mary, xxviii, 164
Stanhope, Lady Hester: xxiii, 34, 37-46
  birth, 37
  childhood and early life, 37-9

death, 46
departure from England, 39
final years, 46
life abroad, 39-46
political connections, 37-8
relationship with Lord
 Byron, 39
servants, 42-3, 46
shipwrecked, 40-1
style of dress, 41-2, 44
Suffrage, 87, 102-103
Sykes, Ella Constance, xviii, xix, xxv

Transport, xxiii, 16, 29, 157
 bath chair, 14, 17, 23
 bicycle, xviii, xxi, xxii, xxiii, 29, 81, 83-7, 165
  see also Annie Londonderry
 boat, 18, 38-40, 72, 108, 124, 129, 148, 156
 camel, xviii, 68, 126, 137
 canoe, 16, 59, 173
 carriage, xxv, 36, 147
 on foot, 3, 11, 16, 53, 107, 134, xviii
 on horseback, xii, xviii, xx, xxi, 9, 20-2, 53, 137, 151, 154
 palanquin, xxiv, 157
 train, xviii, 82, 84-5, 93, 99, 168
 ship xviii, xxvi, 5, 40-1, 49-50, 55-6, 58, 84, 94-6, 123-5, 143, 148-9, 155, 178
 sledge, xviii-xi
Travelling attire, xii, xix-xxiii
 see also Fashion.

Victoria, Queen of the United Kingdom of Great Britain and Ireland, 138-39, 144, 165
Violence,
 see danger.
Verne, Jules, 37, 88, 91, 96, 99
 See also Nellie Bly.

216